Stepping
As
Before

Richard V. Bovbjerg

Linda and Duane
dear friends
Dick

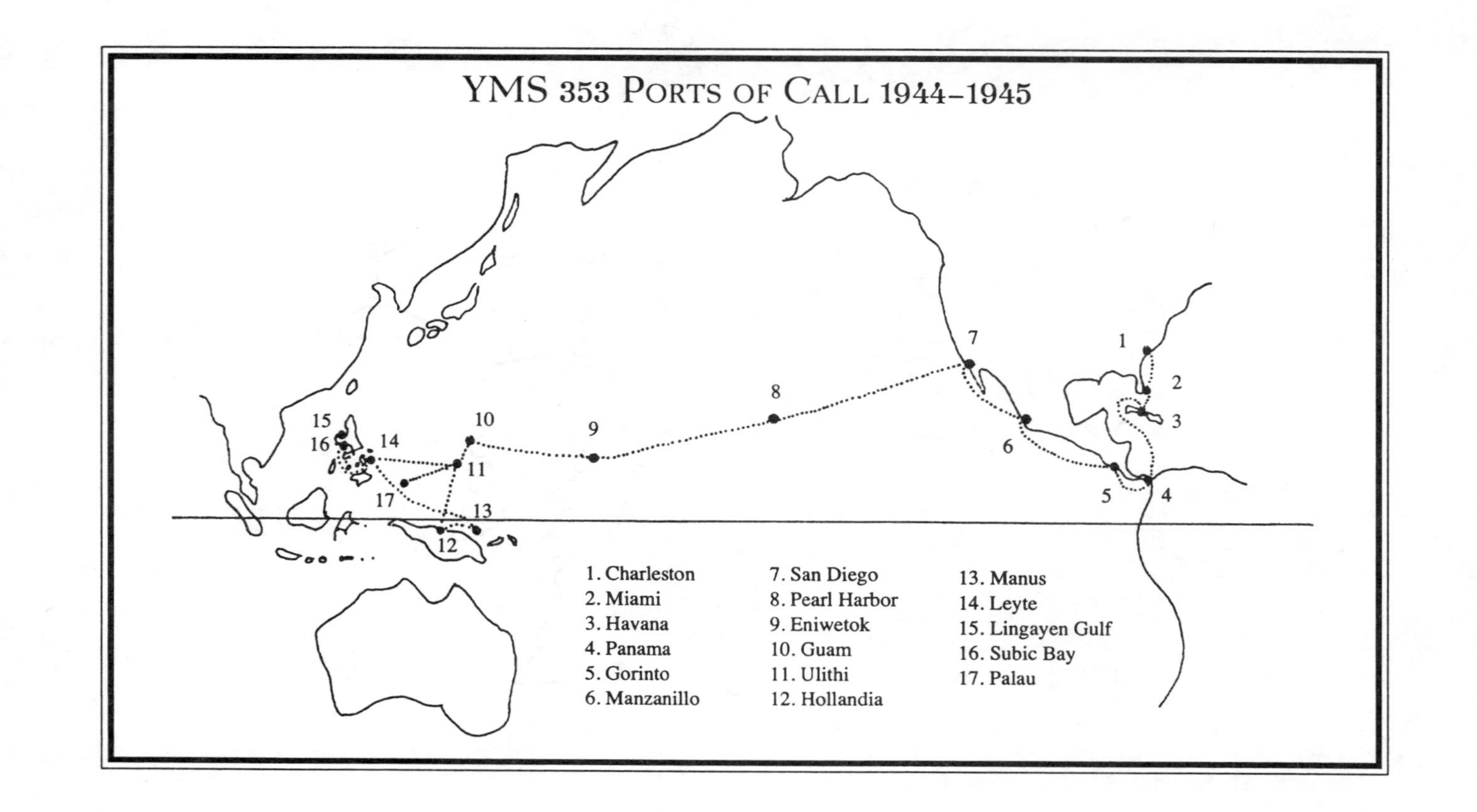
YMS 353 Ports of Call 1944–1945
1
2
3
4
5
6
7
8
9
10
11
12
13
14
15
16
17
1. Charleston
2. Miami
3. Havana
4. Panama
5. Gorinto
6. Manzanillo
7. San Diego
8. Pearl Harbor
9. Eniwetok
10. Guam
11. Ulithi
12. Hollandia
13. Manus
14. Leyte
15. Lingayen Gulf
16. Subic Bay
17. Palau

Steaming as Before

Richard V. Bovbjerg

International Scholars Publications

San Francisco • London • Bethesda

Library of Congress Cataloging-in-Publication Data

Bovbjerg, Richard V.
Steaming as before : wartime minesweeping, etc. / Richard V. Bovbjerg
p. cm.
ISBN 1-883255-12-0 : $59.95. — ISBN 1-883255-05-8 (pbk.) : $39.95
1. Bovbjerg, Richard V. 2. World War, 1939-1945—Naval operations, American. 3. World War, 1939-1945—Personal narratives, American. 4. United States. Navy—Biography. 5. Seamen—United States—Biography. 6. Mine sweepers—United States—History—20th century. I. Title
D773,B68 1994
940.54'5973'092—dc20

94-18875
CIP

Cover and Text Design by Diane Spencer Hume, San Francisco

Published by
International Schalars Publications
7831 Woodmont Ave., #345
Bethesda, MD 20814
Phone: (301) 654-7414
Fax: (301) 654- 7336

To that crew of the YMS 353. Thanks.

Contents

Preface

These pages were written for the children. Anyone else is cordially invited to eavesdrop.

You lads heard sea stories early and you knew who it was that won World War II: Then came the time you had your turn. "Come on Dad, 1941, there were people then?" Now this is my chance to relive some of those tales with you, put things together, and tell it like it was. You four guys with your different and splendid talents, these tales are for you.

This year is the 50th since I joined the Navy; in October, 1941 I put on the bell bottom trousers and white hat. Our country would soon be shocked by the bombing of Pearl Harbor; I was stationed then in Chelsea Navy Hospital in Massachusetts. No one knew that we would have four war years.

I am going to recall only the last two years, year and a half really, starting when I was given command of that minesweeper.

The narrative is rambling and anecdotal. This is not a scholarly study, though I have as you know read the major works on the Navy in World War II. And I have Hydrographic Office charts of the areas in which I served; these were wonderfully useful. This stuff is vaguely chronological, but more topical. It's done from memory. We were allowed no diaries or journals, but fortunately my memory is clear on the episodes reported. The National Archives and the Historical Center in the Washington, D.C. Navy Yard had the ship's logs which were very reaffirming. But, as you will see, log entries are terse, bare facts; no stories there.

I feel that I owe you some explanation of the structure and style you will find. Not one line is fictional; there is no plot, no intricate play of characters. There are, there must be, mistakes. Memory is surely selective. And remember, these episodes are recalled by the Captain; the same events would be recalled by a crew member but his recollections would be so very different.

These sea stories range from terrifying to dull. The tone and feelings will be hard for you to grasp; there is an extraordinary generation gap, a generation that went from years of depression into total war. The social differences from your generation are unbelievable; that will be clear.

Some threads will emerge. 1.—The title, "Steaming as Before," epitomizes the entire year and a half of anonymous mundanity. 2.—You will see the metamorphosis of a dumb college kid officer to skipper of the YMS 353; astonishing. The naval war was run by professionals but done by amateurs—citizen sailors. How was that possible? 3.—You will also see the "The Navy Way" of that time, pluses and minuses, all through these pages. 4.—And lastly our experiences were those of thousands of small ships at that time; it cannot escape you how special were the lives and problems on these fragile, little tubs on such perilous seas. Regular Navy will not believe this!

Nowhere have I used names of the guys on the ship. I cannot forget those characters but I have not retained many names. I did not want to slight any one of them; all of them mattered so much.

There will be a lot of personal probing on my part of all this war stuff. I was an introspective man then and I have had reflections since. These could be interesting to you; they were very much of my life, as important as some of the events. Hope it does not bore you and you might even get some insights to the quirks that you have seen in your father.

Again, don't expect a literary gem! Page after page I am just talking with you about the events and feelings. So the style and tone are deliberately conversational. No dissertation this.

Hey, you guys, doing this has given me so much pleasure. It has become a mission. I have thought about it and thought about it, one memory giving rise to others. Doing this was an end in itself—so much of life is. And the doing of it has brought me so close to you four. It is a bonus if you too can enjoy some of it. Hope you do!

Richard V. Bovbjerg
Iowa City, Iowa
December, 1991

"For Duty as Commanding Officer"

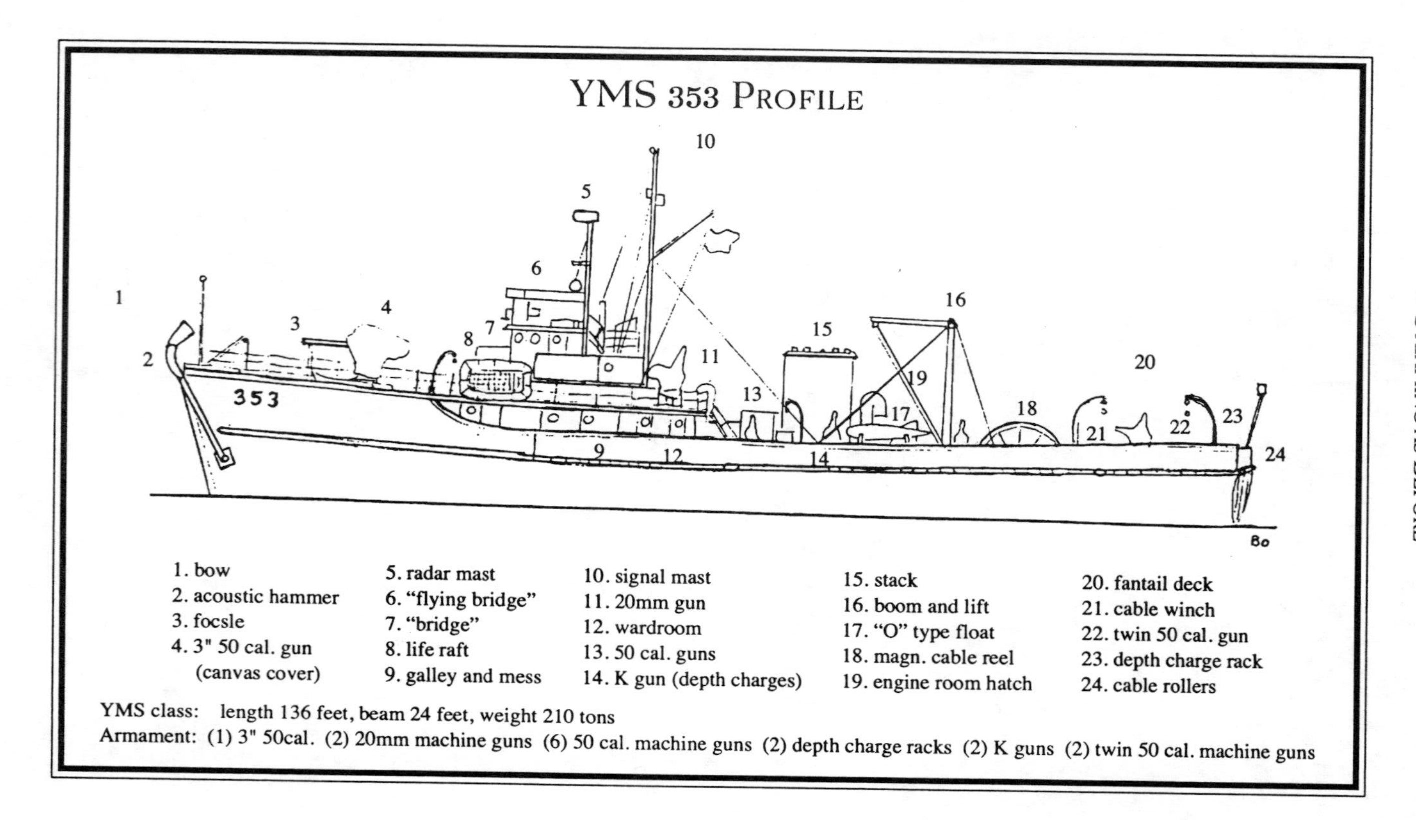
YMS 353 PROFILE
353
Bo
1. bow
2. acoustic hammer
3. focsle
4. 3" 50 cal. gun (canvas cover)
5. radar mast
6. "flying bridge"
7. "bridge"
8. life raft
9. galley and mess
10. signal mast
11. 20mm gun
12. wardroom
13. 50 cal. guns
14. K gun (depth charges)
15. stack
16. boom and lift
17. "O" type float
18. magn. cable reel
19. engine room hatch
20. fantail deck
21. cable winch
22. twin 50 cal. gun
23. depth charge rack
24. cable rollers
YMS class: length 136 feet, beam 24 feet, weight 210 tons
Armament: (1) 3" 50cal. (2) 20mm machine guns (6) 50 cal. machine guns (2) depth charge racks (2) K guns (2) twin 50 cal. machine guns

1. "For Duty as a Commanding Officer"

NAVY DEPARTMENT
237072Bureau of Naval Personnel
Pers-3137Washington, D.C.
VEH-66 May, 1944

From: The Chief of Naval Personnel.
To:Lieutenant (jg) Richard V. Bovbjerg D-V(S) USNR
Via: CO.
Subject: Change of Duty.

1. You are hereby detached from duty on board, and from such other duty as may have been assigned you; will proceed to the U.S.S. YMS 353 and upon arrival report to your immediate superior in command if present, otherwise by despatch, for duty as commanding officer of that vessel.
2. Travel by commercial air as directed.

CC: CESF
CO USS YMS 353 signed

Our Yeoman handed me the letter; as the ship's clerk he opened and filed all routine Navy correspondence. We'd just returned from an uneventful five days of minesweeping off Chesapeake Bay and tied up at the section base, Little Creek, Virginia outside of Norfolk. It was a beautiful spring day and the mooring lines were creaking and groaning with the gentle surging of the ship. The Captain had gone ashore for a drink at the club. As the executive officer, I was left with the chores of resupply, getting ready for the next patrol. The yeoman grinned and shook my hand. "Congratulations Lieutenant."

Alone in the wardroom, I read and reread the letter, and then the last line, over and over: "for duty as a commanding officer of that vessel." I was in shock. Stunned. I had only been at sea one year, and I did not want to be a commanding officer. But the orders where there, and one does not say no to the Navy.

My first impulse was to call my wife. Diana worked at the Triple A desk in Norfolk, the auto club. So I'd hitch a ride home and we'd go to the club for dinner at NOB, the Naval Operating Base, a huge naval base in Norfolk. And I'd go to the Operations Office and find out where the YMS 353 was. My guess was Italy. We had just finished sweeping for some huge convoys, presumably heading for Italy. They included YMS's.

Diana and I went to NOB in our little Willys-Knight car. First we went to Operations. I was kicked around by junior officers but the words, "For duty as commanding officer" could not be brushed off and I finally got into the War Room where there was a multi-floor map of the seven seas that showed the locations of all the ships of the U.S. Navy. The 353 was not in Italy. She was in Mayport, Florida. Her skipper had heart problems and she was pulled out from the Italian invasion—so the story went. The sweeps had had a tough time in Italy; we lost some.

Out in the car, I told Diana the news. She put her arms around me and held me close.

"You're sure then?" she gasped.

"Yes, yes."

I told her of my skipper's dream, of his ship being sent to Mayport. It seems that he came through there on the shake down cruise and never forgot the wonderful officer's club with the gambling casino, and the beach and the blue water. I hated the very idea of going back and telling him my news.

As we walked over to the officer's club our feet weren't touching the ground. Diana held onto my right arm, leaning on me and whispering in my ear. I looked up in the darkness and there was a four striper, a Captain. Too late to free my arm for a salute.

"Young man!"

I wheeled around.

"Sir!" I made a stiff, snappy salute.

"Who is this?"He gestured toward Diana.

"My wife, Sir."

"You do know the courtesies?"

"Yes Sir."

"What's your ship?"

"Who's the Exec.?"

"I am Sir."

He stared at me and snarled.

"Watch it mister."

He strode off shaking his head, wondering about the damned "feather merchants" (reservists) and what would happen to the U.S. Navy. Diana was convulsing silently against a tree, in one of her pixie streaks.

At the club we had the full $2.50 dinner after two bit Canadian Clubs on the rocks, several courses served by white jacketed waiters with carts. We held hands under the table. Then we picked up a couple of bottles at the package store and went home.

Next day I told the skipper. He went beet red—speechless, "Mayport!" He stumbled down the pier to the club and then spent the next two days in the sack nursing a hang over. The yeoman and I compiled the letter detaching me and directing me "to proceed in accordance with basic orders."

Our entire belongings would fit in the back of the car. The Navy gave me five days, to count as leave, and a book of gas ration stamps. In Florida I was reimbursed $54.24 for the mileage from Virginia to Florida. We coaxed the old car down Highway 1—saw the southeast coast.

In Savannah, Georgia, we got one of the last rooms at a hotel. We were beat. At ten o'clock the phone rang.

"This is the House Detective, sir. You will have to entertain your lady friend in the lobby."

"What! She's my wife."

"The register does not show that, sir. Come down and straighten it out please."

"No! I told them at the desk, and I'm ready for bed."

"I will have to come up to visit you, sir." I thought Diana would burst.

When he knocked at the door I was utterly embarrassed, and so was he. Not Diana. She was sitting on the bed with all her clothes on, but had pulled the sheet up to her chin, cowering and grinning with such a terribly little girl face—such sweet, honey and silk, with ginger. I found an I.D. card of hers in her purse. It was for the Naval Operating Base in Norfolk, as an officer's wife. He bowed and left. Diana burst into uncontrollable laughter.

It took us two days to reach Florida. We'd been married two years but were still newlyweds and a long trip was heavenly and I needed time to sort out and answer the questions that unsettled me. Having Diana to talk to was perfect. Could I be a commanding officer? Why now? Why me and not my skipper?

At the University of Chicago I'd been a flaming pacifist. We burned piles of wooden guns. We took the Oxford oath never to bear arms. I told my Dad on graduation, "I will not be drafted." He smiled. I had a college deferment but I had a very low draft number. The draft board had instructed me to stand by for induction after I graduated in June, 1941. Twenty-one years old, good health, lean, athletic, lettered in both high school and college.

I went to all the recruiting stations in Chicago; I chose the Navy. They would give me a third-class Pharmacist's Mate rating in the medical corps based on my degree in zoology; the chemistry and the anatomy and the bacteriology. So I signed up for four years. Diana was angry; the army was only for 18 months. But I knew, and I tried to convince her, that we would be at war and everyone would be in for the duration. I told Dad. He smiled and shook my hand and said, "You will at least mend bodies, not break them."

I put in a year as corpsman, mostly in the Boston recruiting station, giving physical exams where I personally squeezed half the balls of the city. Then came Pearl Harbor and that Christmas Diana came to visit. She was in her last year at Chicago. One night she said, "This is really silly, we might as well get married." It seemed so simple and obvious. We were supremely happy even when I worked nights and she worked days as a waitress at Shrafft's Tea Room in Boston.

A Medical Officer I worked with said, during one of our midnight to eight watches, "You know, you should be commissioned. You are at a job not using your abilities. I'm going to recommend you." He put it through. The Navy needed ensigns. They were losing them very fast and I had to admit I could do more for our cause on a ship; more than in the Post Office Building in Boston. We were at war and the war needed finishing.

The commission came through. One afternoon I went with my papers to a department store. I wore bell-bottom trousers and a white hat when I went in and I came out an officer and a gentleman. I went to two schools. Then in May 1943 I was assigned to the minesweeper in Norfolk, VA. I still looked like a college kid with a big grin; "No warrior he," I can assure you. Was I an officer? Could I stand watches at sea? Of course I was an officer! I had one nice thin gold braid. But the fact is I learned fast. And suddenly I realized that I had been giving orders to the men in my conversations on the bridge. I was a good number four watch officer (we carried four officers).

Then came an amazing year. The skipper got orders to take a new YMS and the executive officer moved up to skipper and I moved up to number three. Meanwhile, we worked in and out of Chesapeake Bay. We went out 100 miles to the edge of the shelf (which is mine laying depth) sweeping for magnetic mines (which is the last mine that the Germans used, actually magnetic-acoustic).

So it went for a whole year—foul weather, fog, storm, ice, snow; heat in the summer—it was a grueling year really. Two months after my first skipper left, the new skipper got a new YMS; the Exec moved up and I became number two! So that summer I was the executive officer—second in command. The crew came to like me; I was someone to talk to, a person who was empathetic. But I wasn't as good as the quarterly report on Ensign Bovbjerg.

The Captain saw the promotion pattern; he was next for a new ship. He showed me the fitness report; I was a paragon of young officers, wise beyond my years, in all respects ready for command. The skipper turned in the next quarterly report: "Ensign Bovbjerg had distinguished himself in combat, a natural perfect ship handler." That report was based on the following incident.

At Christmas of '43 the skipper took a two week leave; I was in command. The second night after the captain had gone we were called out on an emergency anti-sub patrol. The shipping was being attacked and they needed escorts with sound gear and depth charges. It was an awful time. Storm and ice—very, very bad. It was really very tough. and there were no subs sunk! We returned beat but I was satisfied.

I kept telling the captain that he couldn't say those things about me—the nice fitness reports. He said, "I don't care, I'm going to get my ass out of this ass-hole of creation."

Next spring another stupid fitness report, even more glowing, than any of the others. I gave up. Who would believe it in Washington, anyway?

In the days of driving south to the YMS 353, talking with Diana about it, I believe that those reports doomed my old skipper. In Washington, the sudden call for a YMS 353 skipper came in; somebody went to the files and pulled out the file of the top young officer in the minecraft of the Atlantic Fleet, Richard V. Bovbjerg. Fraudulent reports made me a captain of a naval vessel. This thought did not ease my apprehensions.

We rolled south to my confrontation with reality on a new ship. I would be in command, so why worry? After all, I could handle the ship. The old skipper had given me some responsibility. I knew the paper work and I knew the ship. It had been a year of unrelenting tough sea duty, facing subs and mines in very cruel waters.

Maybe I was as good as the Captain. Diana knew me as more the worrier than a warrior. She encouraged me as we drove, and could honestly remind me that "this too shall pass."

What saved me was a very pragmatic dichotomy. My success would depend on two things. Could I run the ship? Of course. Could I command in battle? Who knows? But at this point who cares. There isn't any battle out on the Atlantic now. On the other hand I self-consciously thought of two things. I wanted to have the respect of the crew. I knew about that and I knew that I had to earn it and earn it slowly. It was only by performance that respect would come, nothing else. Two, I wanted a happy ship, and that depended on respect. My God, I'd been at sea one year. Just promoted to Lt. Junior Grade by two months. Was that enough experience?

Mayport was a sleepy base for small craft. A few scattered buildings. It was called a frontier base; the off shore waters were "The Eastern Sea Frontier." Open and windswept, it was separated from the sea only by sand. It opened at right angles into the St. Johns river, a dredged channel. We pulled in; I logged in at the base office. The 353 was out on patrol, due in later.

Diana and I drove around the area (all sand dunes and marsh), and out onto Jacksonville Beach. We saw a sign "Apartment For Rent." Second floor of a 2-unit apartment, right on the beach. Behind us there was a black-top road into Mayport. Then there were dunes and scrub forest to the west —isolated, peaceful, quiet, perfect! We unloaded and were settled in a half an hour and then had our first swim. My God, what a glorious event!

I went back down to the base. Diana was exploring. The YMS 353 pulled in and I took the heaving lines and then the mooring lines and noted how the onshore winds snugged her into the pier. First impressions of the crew were so very important and I resolved that I would force myself to do everything that was needed by myself. I had to show them that not only was I the captain on a piece of paper, but that I was captain in fact. Ship handling was number one test, so I was curious to see how she handled.

That first day I had a chance to look around before the Captain arrived and was glad I had studied the local chart of the harbor. I had looked at the wind and the tide and saw other small craft coming in, maneuvering, tying up. The harbor was a dead-end for tide. There wasn't very much current but the wind was a bitch and the YMS had a high forecastle (the bow part of a ship, pronounced focsle). The bow was like a sail and it would turn away from the wind; the ship was not going sideways but turning as the bow went with the wind. To bring the ship in you had to come in stern closer than the bow if the wind was coming off the sea as it usually did.

The 353 had been brought in smartly; good docking. I was greeted by an officer who asked if I was the new skipper and invited me to come and meet the other officers. We went into the tiny wardroom and crammed into the four places around the little table on a three sided upholstered bench. I was astonished that all three of these guys were big tall men, all over six feet, and all older than I. The Exec even outranked me, but they were very pleasant. The Captain, who had medical trouble, lived off the ship. But he was going to come down the next day and formally turn the ship over to me. We were going out to sea to dump some old ammunition and then change command.

The Captain came aboard about noon. He was a full Lieutenant, very dignified and formal. He was very pleasant and kind. Clearly, he was the Captain in every respect and everybody spoke to him that way. The officers and crew were most respectful and there was no banter here.

We got underway that afternoon. I was still a visitor and had my senses extremely alert. I especially wanted to see how much rudder it took to get out into that damn river which we would have to turn into; it was roaring down out to sea. We rode the river until we were free of the jetties and then we were out on the Atlantic. We dumped the ammo a few miles out and hove to.

The ship was riding on a gentle sea and the crew was assembled in two ranks, one on each side of the fantail, the after end of the ship between the sweeping gear and the depth charges. The Captain and I walked out slowly and the Exec called, "Attention on deck." The Captain said a few words, read his orders to be relieved of command, and introduced me. I read my orders and turned to the former Captain and said, very quietly, "I relieve you, Sir, and good luck." Suddenly the two twenty millimeter cannons let loose with a burst of anti-aircraft fire with tracers. I raised my voice and said, "All Captain's orders and ship's regulations continue in order."

The sky was clear blue and the water was beautiful. Everything was sharp in my eyes. The two of us went back to the bridge and the Exec dismissed the crew. There had been no smiles, all very formal. I asked the officer of the deck to continue at the conn, and to log the change of command at 1525, 19 May 1944.

My predecessor took me aside and said, "I would suggest that you ask me to take her in this last time and dock her. It's rugged if you've never seen it done here before." I grinned at him and said, "No time like now to do it myself. But thanks, it's very gracious of you." He didn't leave the bridge.

When we got in between the jetties at the river mouth I told the officer of the deck, "I relieve you, sir." "Aye, aye, sir." And he gave me the engine revolutions and the course in effect. I slowed it a bit till we were just moving in the current. The left hand turn came so soon and I did not want to miss it. A fresh breeze was coming on shore from the sea behind me, so when we turned to port that would put us to beam of the wind and the high bow would take us to starboard. I had be aware of not getting too close to the rocks on the starboard side of the entrance. More apprehension.

Then a strange thing came over me, something I had experienced before. Ship and man mated. Everything but the ship and I faded and I went into a ship handling trance, a detached, internal focus. My commands were low and slow. I became an automaton, calculating without numbers, thinking of a course, about speed, about the turning circle, river current, wind, tide. Each of these was a factor and they complemented and antagonized each other. Their algebraic sum was a vector that we would follow. It would either shoot us through in the channel or plow us up on the rocks. This asks a lot of one human's mind.

We whipped into the harbor, right down the middle of the channel, and we were in slack tidal water. The wind seemed gentle but I knew I could rely on that later. That test was over. The former Captain smiled and said something about nifty ship handling. Now suddenly the bridge and people came back into focus and I dropped out of my trance. I was weak kneed and made a joke about luck, but doing this myself was the vital first test.

The next test was bringing her alongside without banging the pier; I had seen that so often. All hands were at Special Sea Detail, which was a General Quarters drill, not for shooting, but for coming to port or doing something at sea that required all hands. All eyes looked up at me on the bridge and at the pier; I had thirty-some critics aboard. I knew this and had prepared myself for their concern. This next move was crucial. We neared the pier and turned parallel to it. Time to slow the ship.

"All engines one-third."

"One-third. Aye-aye, sir" was the helmsman's response.

"Steady as she goes." To myself, I said, "Always get half again too close before you stop the engines." This was my over- cautious nature. I knew it and it was one more factor I had to take into account.

More seconds; we closed. "All engines stop."

"All stop, aye aye, Sir."

Now, now back down you fool. No. Drift for a while yet. Then, quietly: "Port back one-half."

"Port back, aye."

"Rudder mid ships."

"Mid ships, sir."

I dreaded the terrible overshoot that the ship takes because of the long time for a backing screw to slow the ship. You don't have brakes. My God, we seemed close! But the stern began pulling in under one backing screw. "All engines stop."

"All stop, aye."

Still too far out, but the heaving line from the fan-tail deck just made it.

Soon the stern line was secured. The bow swung in with the wind and we snugged up precisely. I called for a bow line and a spring line; fenders were put over the sides; we had a tiny bump.

"Secure all main engines. Secure Special Sea Detail. Set the watch. Shall we go below, Captain?" I said. Could he hear the slight tremor in my voice, see the sweat? He quietly went below with me and left the ship immediately.

Log of that afternoon:

> Moored as before. Starboard side to, main deck. US Navy Frontier Base. Mayport, FL. 1400, underway to dump 3 inch powder over the side. 1450, commence dumping. 1500, secure from dumping. 1525, crew mustered at quarters. Captain relieved of his command by Lt. JG Richard V. Bovbjerg, DV USNR. Returned to base. 1625, moored outboard YMS 112, starboard side to USN Frontier Base, Mayport, Florida. 1630 Lt. [And interestingly, originally said, "Captain detached" but the Captain was crossed out and Lt. written above it.] [And down below] Examined and found to be correct, Richard V. Bovbjerg, Lt. JG USNR commanding.

We were moored in the quiet heat; liberty was declared and marked by the departing shouts of the men. The new Captain sat in a deserted wardroom almost crushed by the swift shift from ex-executive officer to commanding officer. I was no longer part of "us;" I was "him." The loneliness was so real even on a tiny, crowded ship. Each man on the ship had his own skills, duties and decisions, but the big ones and the sudden ones were now mine alone to make.

Breaking in a New Captain

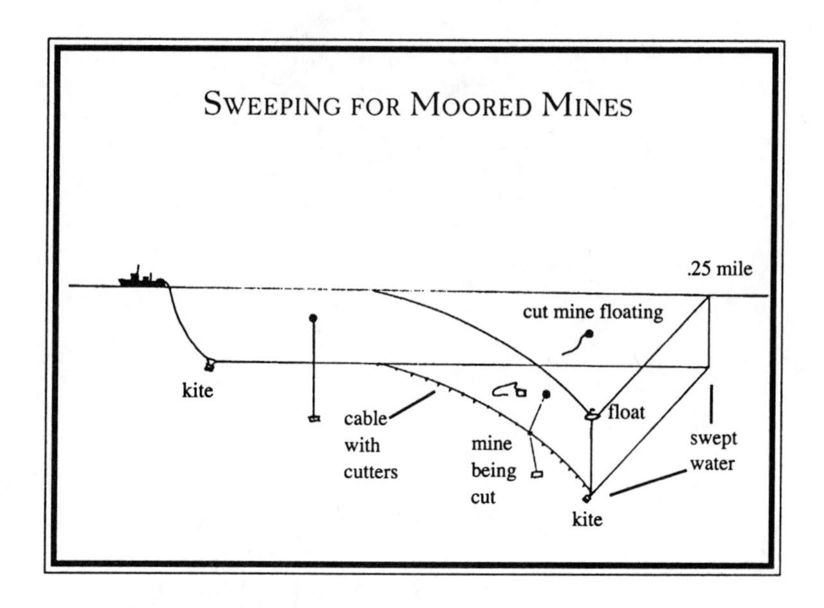
SWEEPING FOR MOORED MINES
.25 mile
cut mine floating
kite
cable with cutters
mine being cut
float
swept water
kite

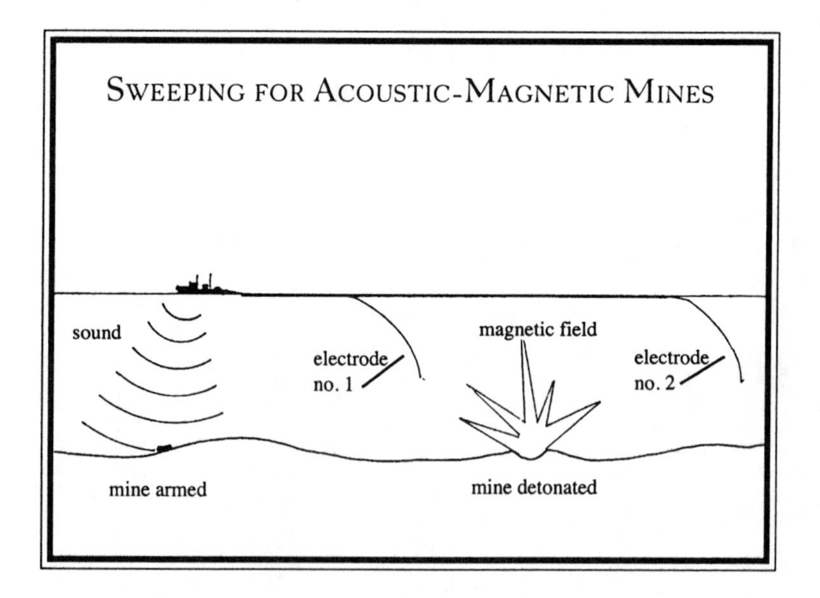
SWEEPING FOR ACOUSTIC-MAGNETIC MINES
sound
electrode no. 1
magnetic field
electrode no. 2
mine armed
mine detonated

It was obvious to us that in the early summer of '44 there was a phony war on the Atlantic. The subs were just not here. They weren't attacking our coastal shipping. As a matter of fact, they were being hunted down themselves. The in-shore battle ground was quiet, but we patrolled for subs and swept for mines. Mayport was no longer a combat center. Earlier, the Caribbean, the Gulf of Mexico and the Atlantic Coast were German submarine playgrounds and the number of ships sunk by mines off the St. Johns river and in the Gulf was very high. It had all been very bad, but it had passed. The Coast Guard still ran a horse patrol up and down the beaches but the worry was over. Mostly we worked days and tied up for the night.

Mayport had not always been as sleepy as I knew it. One of the early desperate moves that the Navy made was to appropriate sea-going yachts for coastal patrols. They often commissioned the owners to act as captains of these strangely gray craft. They were called YP's, Yard Patrol boats. We called them Yippees. The men were patriotic yachtsman and some of them must have been useful. At least somebody was out there to sound the alarm. We were so badly equipped for the German attack early in the war. The first year of the war the Atlantic was a battleground; it had been a massacre! The beaches were black with thick tar for hundreds of miles, massive oil spills from tanker after tanker aflame in the night. Hundreds of ships were sunk in the first year.

A Vietnam era writer coined the words, "WWII was the last good war." This writer did not really experience WWII. The first year we worried about surviving as a nation! Hitler came close. "Good war?" And that was before we knew of the tens of millions dead in Europe and Russia or of *Kamikaze* destruction in the Pacific. "Good war?" Atom bombs = good war?

I don't think these Yippies accomplished a lot. They carried a radio and a crew of two or three or six, whatever the size of the boat. They would have a machine gun and flares, but they didn't have any radar or sonar. The vessels themselves were not commissioned, so they didn't have a Commanding Officer; they had an Officer in Charge.

One such captain reached fame and dismissal from the service through his stupidity. He talked too much, he called in all the time and never knew what to do. Once he was told testily to return to base and did not arrive. Hours went by. They called him, asking why he wasn't back.

He answered, "I don't understand it. I've been heading 090 for four hours and have not sighted land yet."

They answered, "Try 270 for five hours."

Another skipper was the opposite, a warrior. One night he was too close to shore, and the boat rammed into an off shore sandbar. They were knocked off their feet. He hollered, "Hey we've hit a sub. Back 'er up and ram 'er again." And they did. They rammed their way across the sandbar into the breaking waves, rushing to the sandy beach. He was grounded in a pool of water between the sandbar and the beach. The next day a tug pulled him out at the highest tide. He did not last long. The Yippies faded but there were still some in the harbor there with us.

For us it was routine sweeping and patrol each day. But we would be called away for a few days at a time and patrol other river mouths where heavy shipping was leaving. We still swept for any ship movement whatever. At Brunswick, Georgia they built Liberty ships. When they were going to send these out to sea we went up and swept. One new ship had been torpedoed right outside of the river. We also swept off Savannah, the big harbor there was very busy.

These were great liberty towns. Upriver, safe mooring, old fashioned riverfronts. The officers always found the glittering hotel bars, and the crew found the noisy riverfront dives. All hands would pour aboard later in the night. We'd be off at first light in the morning.

For a two week period we hosted a dozen or so ROTC students from Georgia Tech. This gave a dozen home leaves for some crew members. It was good for the ship and was a great two week experience for the ROTC midshipmen.

The students stood watch and slept in the crew's quarters. They were kind of sick, but very impressed with a "real warship." I slowly got to know the officers and crew. We were at sea for days on end. This was a nice dress rehearsal for our inevitable removal from these calm waters. And the confidence in the crew did grow; it grew both ways. We were in an out of tricky rivers and strange dockings. The ship handling never failed. There was nothing dashing, but solid work. I got to know more of the crew and to respect their abilities. We four officers grew closer.

I thought a lot about ship handling. All my life I had rowed and paddled boats and canoes. I knew a very basic fact, that handling a boat or a ship was all a matter of skid, controlled skid. After all, a boat skids all the time; no wheels to turn; no brakes to stop. Just skidding. The other officers were more experienced on wheels, bikes and cars. This was very different and we talked about it.

Just think of how much it took to stop or turn a ship. This was the skidding part. Speed times mass equals momentum. On the ship, we could barely turn in the length of two football fields. All was done by voice command, all was done by quiet voice command, very detached.

There was a little open deck on each side of the bridge and I always went out there so I could see. I could look down over the side and then call orders over my shoulder, never touching the throttle or the wheel. So the words must be slow and clear, the communication and response instantaneous. The officer and bridge gang formed an intimacy that rested on a two-way confidence.

A very dramatic event took place on that ROTC cruise. We went into Chesapeake Bay. My old port at Little Creek, Virginia was our destination. We were going to discharge the ROTC guys there. We pulled up from the south. It was pretty, a fabled coast. And how familiar those capes looked; must be twenty miles apart. The buoyed channel down through the middle was through our own minefields. There were a lot of buoys just outside the minefields marking sunken ships that had been hit by German torpedoes or mines.

The sun was just setting over Chesapeake Bay when we went in. It would be dark before we got to Little Creek. The Exec tried carefully to ask if maybe we should anchor in the Bay, and come in in the morning.

"No," I said, "Let's go, tie up, and relax."

The officers, unaware of the time I had spent at Norfolk, were a little bit bewildered at this rash move. This was a new bay to them and at night it was alarming. You could barely see land because the of the marshy swamp lands around Chesapeake Bay. There's not much in the way of landmarks. There were no lights, as this was wartime. We did have radar and could see that the shore was so many miles off and that we were making progress to the west. I'd come in here, in that year, in rain, in dense fog and storms. This was a beautiful clear summer night and as we approached you could make out the long stone jetties through the field glasses in the starlight.

We called Special Sea Detail for coming into port. The deck gang was rousting out the mooring lines, and there was scurrying around on the deck. The signalman was on the bridge standing by. The Exec and I were on the bridge and I took the Conn. Time to slow.

"All ahead one-half."

"All ahead one-half, sir."

Then off to starboard, my old enemy, the Capes Railroad Ferry. She went back and forth with a huge tug and a string of long barges filled with railroad freight cars. If she beat us in, she had right of way; we'd be forever getting in, because she filled the channel and went in so slowly. So I said, "All ahead standard."

"Ahead standard, sir." We zoomed all of a sudden; and we beat them by crossing their bow, way ahead. Lots of nervous glances my way. We dropped speed again when we came to the jetties. We cleanly hit in the middle of the channel, right between the jetties, and slowly cruised in. The rocks seemed awfully formidable that night, but I'd done it so often.

The signal tower was raised. I asked the signalman to flash 'em the old "dah-dih-dah-dih-dah" (means a request to signal), and ask for berthing instructions. They flashed back the signal. Then our signalman asked me, after a long pause, "Mr. Bovbjerg, do you know what this signal means, Baker-Three-Sugar?"

I delayed my answer. "Uh, send a wilco." (That means will comply.) And then I said, "That must mean the second pier, third berth, south side." I asked the Exec to tell the deck gang that we were coming in portside to. I intended to back in because two backing screws gave more control. The blunt stern was easier to guide, in current or wind, than the sharp bow acting as a vane. All the lines and fenders went out on the port side and there was a flurry of action on deck—this was going to be parallel parking in a very close parking lot without wheels or brakes.

We slowed, and then the harbor unfolded before us. There was a maze of dim red lights, one on the mast of every ship. The railroad ferry hissed and groaned in behind us.

Then my state of alertness came over me and the world faded; I remembered this drill so well; had done this maneuver many times. There was a maze of piers, like fingers that pointed to a large turning basin. We crept into that basin to turn. My orders were so clear because there was dead silence on the bridge.

"All engines stop."

"All engines stop, sir."

We started coasting.

"Hard right rudder."

"Hard right, sir." This would swing us right while we still had some way on, some speed. Slowly we lost that speed as we turned.

Before we stopped, and still swinging, I called for the, "Rudder midship." From here on turning was best done with the screws. "Starboard back one-half." This would swing us more until we were dead in the water. Now the stern faced directly between piers A and B; remember, I knew which they were. We stopped all engines and sat there poised.

Then, "All back one-third."

"Back one-third, sir." We crept slowly back into this very narrow channel between two piers, often with two ships nested together. They seemed so damn close. They were! There was talk between the guys on their crew and our deck crew standing by. The water seemed confused, slapping between the ships. Then I could see that one ship, in the third berth.

"All stop."

"All stop, sir."

I hollered out to the ship where we where going, "Can you take our lines?"

"Okay, sir."

"All ahead one-third." This was our brake; we stopped.

"All stop. Away all lines." We were only a few feet off the other ship. The lines were tossed over and we snugged in.

The crew and the officers were amazed. They looked up at the bridge and I stared back aloofly; only I knew the falsity of that claim to fame. They did not know how many times I had done this on my other ship. It was like smooth silk, but a scam. After this, more of the crew started calling me "Captain" instead of "Mr." or "Lieutenant."

There was a sort of spastic gaiety when I got down to the wardroom. They were all congratulating me on this marvelous seamanship, and I told them, "Well, gee whiz, this was where I had duty for a year. I know this port." But that one night was a watershed for my status.

Back down in Mayport we were met by another assignment which was very strange and very exciting. Go to sea and pick up a passenger from a tanker which was headed north. This tanker had a civilian engineer aboard, some kind of an expert on diesels. The tanker would ride the Gulf Stream coming north. The Gulf Stream was a three, four knot current, and this tanker could get a quarter more speed. The southerly counter current was inside; so from shore, all you saw were ships heading south. We headed out for our rendezvous at a specific latitude and longitude.

It was beautiful out there that day. This was Florida sailing at its best. The seas were not high, just choppy. No huge swells, or so it seemed. And this was a geography lesson; I could not wait. I 'd never seen the Gulf Stream up off Norfolk. Thrilling. We were actually going to be in this very special water.

The color of that huge river of water from the tropics was different. We came up on it and could see it with the glasses. It was pale clear blue. Strikingly different from the water inside of it. We steamed slowly, right into it at right angles. We could stand on the bridge and look over the side and say, "Almost, almost, now, now we're in the Gulf Stream." It was that sharp. And at that time the bow swung north dramatically. The helmsman struggled with the wheel. This was a very powerful force. We could see that we were in the Gulf Stream, and feel it simultaneously.

We radioed back to our base that we were in position. They had contact with the tanker, and told us we had a little wait. We'd left pretty early as it was no place for a tanker to sit alone out there. In that wait we went back shoreward of the current. I recall two feelings. I relished our being all alone out there on the sea with nothing to do. It seemed like a holiday. Not a sight of shore or any other ships.

But I also had to figure out how we were going to do this transfer. How big was this tanker? We didn't know. One thing we surely could do, we could lower our little boat, our little twelve foot wherry and row over. That would be tough though, and very time consuming. Again, we didn't want a tanker sitting absolutely quiet for any length of time. Apprehension crept in.

I talked to the bosun; he was experienced. And we decided to try a Chinese landing that I knew only from the manual. Two ships meet bow to bow, and then the passenger jumps from one to the other. It had to be timed perfectly. The bosun would not have bet on me, I'm sure. But he was most proper, and got his crew together. He put all kinds of heavy fenders on the bow. We were going to meet their bow just off dead center ahead, so that if we did hit we'd be glancing.

Then we saw the sticks, the tops of the masts of the tanker to the south. Then we saw the superstructure, and then the hull. She was only a few miles away. We talked by flashing light. They agreed to a Chinese landing. But it was our responsibility as the most maneuverable vessel; she would lie dead in the water and we would do all the maneuvering.

She seemed to be plunging up and down. I didn't know there was that kind of swell! It had seemed so nice and smooth to us. I could just see us being ground up and the passenger ground meat. There was no danger to the ship; we weren't going to sink, but there could be an awful mash. We went to Special Sea Detail, our best crew on the helm and throttle. I briefed them on the drill. We could not have been more ready.

The whole corps of critics was out on deck watching. This was not something any of them had seen. The tanker seemed like an island! By good luck her bow and our high bow seemed to be about level. They were loaded deep. That was the number one requirement. We had to be somewhat alike in height, or there'd be a hell of a step.

We approached very slowly. She stopped. And we could see she was surging in an exaggeratedly slow motion: plunge, rise and fall; then she seemed to hang poised, and slowly roll over to one side. That was where the bow to bow was good because the rolling wouldn't matter quite so much. She just surged in a giant swell, even dead in the water. We could not feel that deep surge. We got closer, maybe within twenty feet, and I sat there for a minute, and watched, trying to get the kind of rhythm to expect.

The passenger had a life jacket on. He was on the outside of the lifeline on the bow of the tanker, leaning out, clutching the line. We eased in closer, finally, just a few feet apart. Here were these two terrifying masses loose on the sea, one surging and the other hovering. The water between us boiled in compression. It was dead quiet, just the remote sounds of slow engines. I moved in at the peak of her lift, and when the tanker bow was high above us, but slowly starting down, we moved right in, almost touching. Their deck slowly dropped by ours, and at that exact moment the guy just stepped over. He was grabbed by our guys. He slipped out of the life jacket and threw it back down on the dropping tanker. We slowly began a retreat and the tanker ever so slowly picked up speed again. There was a certain joking and thanks between the officers and the crews of the two ships. Our new friend went below with me for a cup of coffee. We collapsed; we were both extremely relieved.

Mines and minesweeping were my life in that unpleasantness of the forties.

The Germans developed influence mines but the Japanese used only the old fashioned contact mines anchored by cables to the bottom. The Germans worked out three types of influence mines, and we copied them instantly for our own mine warfare. They were set off by three different influences: acoustic, magnetic and pressure. No contact with a ship was needed; the mines were laid on the bottom. Even at many fathoms their explosions could crack the hull of large ships and disintegrate smaller ones. The pressure mine came late in the war. We filled Tokyo Harbor with them.

I had spent the year before I came to the 353 sweeping off the Virginia Capes. The Germans mined the track to Chesapeake Bay. We swept for acoustic/magnetic mines which were first armed, or cocked, by the sounds of ships' screws and engines, then detonated by the magnetic field set up by the mass of steel in a ship's hull and machinery. This involved a very delicate wiring job in the mine and a fiendish device which could be stimulated and move a click, like a ratchet, without setting off the explosion until a certain number of clicks piled up. Twenty ships—boom!

These outlandish, mindless weapons kept two minesweepers in day and night operation, 100 miles out and back. As these two came in after five days, they were replaced by two more without missing a stroke.

So how did we sweep them? A wooden-hulled ship could do it, safely pass over the magnetic mine and detonate it astern. But how about all the ship's electrical wiring, the huge engines, guns, and ammo? We had electrical cables wrapped around the ship. They could be energized to neutralize our magnetic field on any heading. We were "degaussed," and then checked by a degaussing station in Chesapeake Bay.

So, if we had wooden hulls, and were degaussed, how did we explode the mines? We had an infernal sound machine on the bow, a jack hammer in a steel drum on the end of a boom that we lowered into the water. This armed the mine as we went over it. We lived in a nerve shattering racket on that tub. Du-du-du-du-du all day and night, eating, sleeping, surviving.

Astern, we hauled two long copper cables; they were kept up by flexible floats. Each ended in a long, bare cable dangling down, some distance apart. When we pulsed positive and negative charges out these cables, we created a huge battery in sea water, enough to detonate a mine (hopefully one-quarter mile astern). That explosion would shake up a YMS but not sink it. The juice was 3,000 amps from huge, straight-8, diesel generators, working in alternate pulses.We always swept with two ships abeam making a wider swath, tough in the North Atlantic winter storms. We had to keep perfect station on each other, distance apart and speed, even before radar. This was gruelling.

Whenever our sweep gear was out we were ducks sitting on a pond, targets. Our speed was greatly reduced and our tails hanging out astern made us awkward to handle. When there were two of us or perhaps six sweeps in echelon then ship handling became excruciating. If we made a sharp turn or stopped, we could cut our tail or foul our screws with the cable. And we could not make a sonar search for subs with the acoustical hammer going; we were deaf to subs. To aircraft we were slow, unmaneuverable targets. Vulnerable. Hellish.

By the summer of '44 the Germans had pulled the submarine fleet back to European waters, and a large fraction of that fleet had been sunk. Just the same, we had to sweep the approaches to those major harbors of the Southeast coast. And once in a while we swept for moored contact mines, just for the drill and the inevitable day we would face Japanese minefields.

Our enemy in the Pacific used contact mines, mines moored to the bottom and floating up below the surface; these mines explode when struck by a ship. World War II contact mines were basically World War I types of the North Sea mine fields. These big metal spheres with lead horns had several hundred pounds of TNT (I don't remember exactly). When the ship struck a horn on the mine the horn was bent, a glass tube of acid broke, the acid spread and it heated the booster; the booster detonated and the TNT exploded—fast. Even if a ship just touched a mine, it would roll along on its circumference until it hit a horn which would break and blow the mine. These mines were moored by a cable to an anchor and they contained enough air to float; so if the cable were ten fathoms in twelve fathoms of water the mine was two fathoms deep.

To sweep moored mines we used what the Navy called an "O type" sweep. This was named after a British ship, *Oropesa*. We let out four hundred yards of cable, or something like that. At the end of this long cable was a big float that rode the surface; attached to the cable below the float was a kite, a box of metal vanes on an angle. That pulled the cable down to the sweeping depth, say two fathoms. Just astern of the ship, at the near end of the long cable, another kite, or depressor, pulled the cable down to two fathoms. So the cable now was hundreds of yards long and all parallel to the surface, down two fathoms. But the kite on the float also pulled outward as well as down. This pulled the cable out and made a long catenary, a sickle curve in the cable; so we would sweep a hundred yards to the side, something like that. This was the width of the sweep.

Clamped on the cable were a bunch of V-shaped cutters, case hardened steel. When the mine cable brushed against this sweep cable, they would scrape by each other until the mine cable was caught in the cutter. The force of the pull (and the YMS had to be very powerful) dragged that mine cable into the cutters in a shearing way and cut it free. Then the buoyant mine popped up to the surface. About a third of it was out of water and it was a sinister black object. We destroyed them with gunfire. We would get a quarter mile away and have target practice, sink them if we put holes in it. But if a horn were hit, KA-BOOM; we had shrapnel all over hell.

All of this has been oversimplified and is of course obsolete.

A crazy piece of gossip was going around then; a diabolical plan had been drawn up by the U.S. for damming the Gulf Stream and diverting the current out into the Central Atlantic. This was feasible only if you really wanted to believe it. The story went that concrete barges would be linked to block the current at the Southern end of Florida, the Straits, *if* the Stream could be diverted, northern Europe would be truly uninhabitable; London and Berlin are the same latitude as Churchill, Hudson Bay in Canada. The whole idea was to bring Hitler to his knees after all of Europe was his. But 200,000 concrete barges? Stop a current a half-mile deep? This story was a sure measure of the desperate mood early in the war.

That summer of '44 we pulled together as a unit. A number of events shaped things. I was accepted. We got to know each other. The Navy carried me, of course. I made the inspections on Saturday morning; signed all the reports; made the decisions.

The Captain was not alone in this breaking in period; most of the crew were rookies. They had been to schools and had skills but not at sea and not with these other men. They acquired discipline, doing the right thing at the right time, always and intuitively. This was a personal thing but hinged on faith in shipmates.

The men knew when I was angry, because it showed in an intensity in my speech and look. But I did not shout. I'd seen that and did not like it.

I managed to work out what so many people have worked out, that I was going to compliment in public; go out of my way, on the bridge or on deck, to compliment the crew. I'd reserve rebukes for private. It worked. Young as I was, it worked. That was sensible, looking back on it.

I had many bad times, in private. My misgivings still haunted me. I always had a little bottle of paregoric with me, a half-spoon of it to settle the gut. One key problem I could not solve. We had some chief petty officers aboard. We really didn't rate such chiefs, but they were put on a new ship to train the crew. There really wasn't enough for them to do on a little ship. They were the top petty officers; they were regular Navy, older than I was, lots of sea duty. They did their job. But they could not accept me, and it showed. That was the stuff of nightmares, but I could not face them about it.

Maybe this was where my having been an enlisted man hurt me. I really looked up to a chief, that's the top rank for a swabbie. I had been a swabbie. I also realized that they saw me as a superior designated by the Navy, and that put us in a funny position. They always said, "Sir." But I was different from many of them in a very special way. I had an open mind; I learned very fast. I worked at it; I was a student. This had been my experience, it was my home life. These guys were highly trained. The chiefs, often very bright, ran the crew and were buffers between crew and officers. The Navy way, a very good way. And the Navy did well to assign them to us during our first months.

We had some great liberties in the summer of '44. Diana and I had the little car, and we had a few gas stamps. I guess we had a couple of gallons a week. We visited Jacksonville and the old fort down at St. Augustine. We ate in restaurants. I never did that in college, neither did Diana. We were much too poor. I used to call Diana in Chicago and ask her out on a date. That meant a walk in Jackson Park. We'd stop along the way and for a nickel we could buy a small bunch of carrots, and that's what we ate.

There were great liberties in other ports. We had some good ones in Brunswick in Georgia. One thing sailors are famous for, enjoying a good night on shore—Liberty.

One night two of our motor machinist mates stole a freight yard engine, a diesel locomotive. After all, they were diesel mechanics. These guys always wanted to be engineers and it seemed like the thing to do. So they put the thing in gear and it rumbled around the freight yard, got shot around on different tracks, depending on how the switches were aligned. And then they got out on the main line in the city, and they went by streets where the gates would drop and the bells would clang at the crossings. Pretty soon squad cars began blocking the tracks. They put it in reverse and went back. Finally they abandoned ship and raced through the early morning in the back ends of the town.

They missed the ship. We took off. They were listed as away over leave, very serious. I knew nothing. I really did not know where they were; this story didn't come out till I talked with them later. I talked with the sheriff, I knew they'd show up. Well they sneaked in after hitchhiking, a very long way to Mayport. Anybody who picked them up must have known they were AWOL.

The Navy said that aboard ship, it's Navy jurisdiction. So I put them on report, and held a captain's mast. A captain's mast is a personal hearing by the captain (in the old days actually before the mast), and the punishment was assigned by me, with a lot of leeway. This was an incredible system of juris prudence. The captain was everything, judge, jury and prosecutor, and hopefully a man of wisdom, but I know it wasn't always the case.

Their missing the ship was very serious. So the report went on their records. It was a serious offense, in one way, but so trivial in another. I restricted them to ship for a long time. After all, it was the liberty ashore that got them in trouble, and that would be no more for weeks. I also read to them from "Rocks and Shoals." These were the Articles for the Government of the Navy. They'd heard these before, but I wanted them to hear the specific sections. I was very sober. They were very sober. The whole ship heard about it; I got the feeling that the men realized that I was serious, and that the punishment was appropriate. But they were also very grateful that I had backed the guys and not turned them over to the sheriff. I was to have many captain's masts. The crew understood that.

There were a couple of guys running down the pier just as we were ready to pull out one day. They jumped on board as we were casting off. They did not miss ship; they were over leave. That meant placing them on report, Captain's mast.

"All right, what happened."

"We think we got married."

"What! How?"

"Well, these two gals found a justice of the peace, and they had a license."

"Are you sure?"

"Yeah, yeah, they did."

"What are their names?" They didn't know. "Didn't you get their names? Didn't you write it down? How about your insurance? Did you turn your insurance over to them?"

"No."

"What are you guys going to do about this?"

Long pause. One of them finally said, "Wait till it blows over."

During this phony war in the summer of '44, Diana was my life ashore. We were so in love, and so terribly young. When we were married we took each day as if I were going to be shipped out tomorrow. We also figured that we would set up house, as long as I was stationed from the States, as though it were forever. Like normal times.

Diana set up house, looked for jobs, from the start. In the year at Boston we ended up living one block from Harvard Yard. And very close to the square and what was then the end of the subway. When I went to school in the Bronx she went home. But then I went back to Boston, for two more months of school there, and we got another apartment back in Cambridge where we felt at home.

One of the things we liked about Cambridge were the student hangouts off the campus. There was a neat sausage house, and some beer joints. There was a marvelous little candy store on Mass Avenue that made brown sugar fudge. It melted in your mouth. I always associate that kind of fudge with that place in those days. It was in many senses just like being students again. We loved it.

In Norfolk, new place, she worked at the AAA auto club. I was at sea most of the time. And we got the car so she could come and get me. And we would have some stolen nights ashore when I didn't have the duty. She came down to the ship sometimes, when I did have the duty in port. We made two more homes, Jacksonville Beach, of which I've spoken, and at old Charleston Town, late in the summer '44.

As captain I stood no watches and we had a phone at the apartment in Jacksonville Beach. When we were in port, I spent most of my nights at home. Diana didn't work. I had been promoted after all to a JG and had a princely salary. I don't recall what it was; it seemed like a lot, especially after having been a sailor. And we had ships' service shopping. We had a cheap car, cheap gasoline. We got around. We loved to talk, walk, and swim. My God, what a break in our lives. At times it was hot, usually there was a sea breeze. We had a four way ventilation in that lovely apartment. A huge living room, all windows, looking east out onto the ocean. Just, maybe, thirty yards from there.

One time it was over a hundred all day. We swam, drank water, and then later in the afternoon we felt sick, light-headed. We stretched out on the floor of the living room under layers of heavy, wet, cool towels, just breathing through them. We slept and made it through the afternoon. This was before air conditioning. That was summer living in those days.

We did a lot of swimming. My . . . it was beautiful,.sometimes big surf, sometimes not. I was on the mens' swimming team at Chicago, and Diana was on the womens.' We had two clubs, a Dolphin and a Tarpon club. And every year the two clubs put on an extravaganza, an aquatic show. That's how we met, at rehearsal. We were good swimmers. Our favorite was body surfing on a mattress cover, a cloth mattress cover that was six feet long, three feet across. We waved it to fill it with air, then quickly tied the end in a knot. Then jump on that damn thing and roar in on those big waves!

We toed over shells as we walked endlessly up and down the beach. This was almost a private beach. Diana would emerge from the sea with the sweet face and form of Botticelli's Venus and I would envelope her with a towel. We watched the sunrise every day.

I explored the small dunes and scrub vegetation. This was so parallel to the Indiana dunes we studied in ecology classes at Chicago. The war, the ship, faded for us, and we returned to our true nature. She was in her element, a tall, slender body, so velvety sweet; wide apart, level blue eyes and smiles. She made a home; we talked of the future. What an antidote to the poison of war, which all seemed so stupid and absurd when we were on the beach, such a terrible emptiness of madness and death.

But then I would drive down the next day to the ship. A job to do, just like always. The crew and I needed the training together; getting ready for combat somewhere. This war wasn't over. It was stupid, but it had to be done. This made the times for lovers more pertinent.

Our last time ashore was in Charleston. She drove up with our belongings, some kitchen stuff, one picture for the wall. We found a place on King street, about two blocks from the battery, at the end of Old Town, looking out on Fort Sumter. Ours was an old slave shack, I swear it was, behind the main house. It was fixed into two-room apartments. Housing was desperate; we were too! We cleaned it up, figured to stay forever or for two days. We'd agreed on that, that we would fix up any place where it looked like I was going to be a while.

I had about a ten minute walk to the Ashley River, where there was a naval facility and a Coast Guard station. I've been back, not long ago; that Coast Guard station is still there. We made mostly daily sweeps and patrols off the entrance to the bay. There was a lot of shipping here, and a Navy yard over on the Cooper River.

Diana tore up the linoleum floor; I helped her. Layer after layer. Apparently, the owner put down more linoleum as the stuff cracked up. We got to the bottom. Then there was a layer of straw on the wood floor and thousands of dead and living cockroaches. We swept them out the door, scrubbed the floor with ammonia, and put down new linoleum. It was home.

We ate pork chops, and taters, corn bread. We walked through the old town at night. Wonderful time, we loved it. Our favorite time was during a light rain, and the old lamps would look dim, and everything had that lovely glistening look. The old houses with the balconies. The trees. Cobbled streets, narrow streets. Old churches. Then the Battery. Gorgeous! Big trees, big seawall, gardens, old guns pointing out.

She waved from the sea wall of the Battery when we steamed out. Sometimes when she had an idea we'd be coming back in that night, she'd be back on the battery, waving. We came close by. The crew would wave to her; she was after all, quite a dish, but a very distant dish. Most of them she'd met aboard when we were tied up in port. "The Captain's Lady."

But of course it ended. In August we went off to war in the Pacific. The phony war was over and we were heading for the real one. In Europe we had invaded Normandy and were closing on Paris. Our stage would be different, but it was the same show. In the Pacific, it was island hopping, and that's where we were headed. Our summer of '44 was over. It was our maturation. It was a time of war, but we were not in combat. I was so grateful for that. Combat would come.

The Tropical Cruise Ship

NORTH SHORE OF CUBA AT DAWN

3. The tropical Cruise Ship

Suddenly, we were on a tropical cruise of Latin America, all expenses paid. It all happened so fast. I knew we wouldn't be long in the country when we got what the Navy called, "Availability," time in the Navy Yard (this time the Navy Yard on the Cooper River in Charleston). The availability was for repairs and replacements and renovation all the way through, drydock, clean, paint; all the machinery was overhauled; we were granted all the latest gear authorized for our class of ship.

Diana and I had a bit of leave. Knowing how long we would be, we took off for Michigan where the folks had the summer camp. This was in late summer, about the end of camp time. We flew most of the way. And we had a few very nice days; there were such fond memories there of my childhood. Then I got a call, a long distance call from the Executive Officer. He obviously had my number there. He didn't say anything; he just whistled, "California, Here I Come." And then he said, "Right away." If anybody had been listening, it sounded like I was supposed to go to California. So we got the next train from Grand Rapids to Chicago and from the train station in Chicago out to Midway Airport and back to Charleston. We folded up our latest little home, crazy place; packed our little car, our Willys. We said our goodbyes the day before we were to sail.

We got our orders to leave on the 20th of August, 1944 in Charleston:

> When in all respects ready for sea, (etc., etc.) you will join the Pacific Fleet in Coco Solo, Canal Zone, Panama.

And there were provisions for our stops in Miami and in Havana, Cuba on the way to the Canal Zone, for replenishing our fuel, water and food.

This message, of course, made this departure from Charleston a bit different from our usual departures when we would make a run out to patrol or sweep and return for supper.

We slowly pulled away from the Ashley River base, away from Coast Guard cutters, minesweepers, patrol boats. In minutes we rounded the tip of Charleston at the Battery and everything was so familiar but so different this time. The Battery was Diana's and my favorite place, a lovely, lovely park at the very tip end of old Charleston. Old guns (which I'm told were never fired in anger on any of the forts), were spaced among the huge trees, lovely gardens and bushes, flowering bushes; all behind a massive sea wall, (high seas in that huge harbor sometimes made huge splashes).

Always, Diana was there to wave at us when we sailed out. It was only a couple of blocks from our place on King Street. There was no wave today. I looked back but the wall was desolate. I knew she would not be there. She had left the night before. But it just didn't seem right not to have Diana waving.

The town faded behind us. We turned our eyes forward but also our minds. We steamed out to sea between the muted menace of Fort Sumter and Fort Moultrie, out into the chop of the open sea, and turned south. It was hard to imagine that it could seem like such an ordinary day, like so many that we'd had there.

So we turned south from Charleston on the inshore southern current, just off the land and we saw only distant landmarks, but such places of memory. We'd had an idyllic couple of months there, Savannah, Brunswick, Mayport, St. Augustine. I was pained to pass Jacksonville Beach where we had spent that incredible time. We neared Miami. And then we had another adventure, of course. Miami, our first stop.

That month in the late summer of '44 was unreal. We were on a cruise; we were on this tropical vacation. A four week cruise of sparkling seas and sunshine and carefree living. We would sail by countless islands, coral keys, and high tropical islands. We had shore visits and nightclub flings in Miami, Havana, Panama City, with stop-overs in the quaint Latin American sea ports of Corinto, Nicaragua and Manzanillo, Mexico.

Miami was an overnight stop for fuel, water, and a few provisions. My memories are vivid but very limited. The city was not memorable. Sort of a big/little town. I wandered around downtown a bit by myself, absolutely unimpressed. But I had known Chicago and Boston and Manhattan.

Coming into the town was memorable. I recall the landfall. The well-marked entry took us from the empty sea where we'd just been galloping down all by ourselves into an especially crowded harbor. Obviously the channel to the downtown piers had been dredged straight as an arrow. This very narrow channel had a long boulevard out to Miami Beach from downtown Miami on the north side and we steamed in at half speed with traffic right alongside us. And suddenly we were not only not at sea, we were practically civilians. The drivers waved, we were so close. The fathometer readings were good; it seemed like we surely should run aground. We were not only close enough to throw a stone, but we were so much in the civilian world it was shocking. No Naval Base this.

The strongest memories are of Miami Beach and Miami's waterfront dives. I had never cracked a coconut. It took forever. I'd never seen palms like this, had never been in the subtropics. Nor had I ever seen such opulent hotels. I remember the "Fountain Blue"—grotesque! Lobby with plaster statues and fountains and incredible chandeliers, swimming pools, beach cabanas, all along. This was not in my kind of childhood. Nothing like it. It was so cluttered. So vulgar and commercial. I understand now it's many times worse; never been back to Miami Beach, although I've been by there many times.

The place burst on us after several days at sea; a bomb so gaudy and noisy. And there was a "what war?" feeling about it—there were lights on. We had money, we took cabs, absolutely new to me. Bars and floor shows. Doesn't it sound like a rather strange impression of the whole city? But that's really all we saw. I had no feeling that folks lived there—just played there.

We took off early the next morning. The city was so vulgar, and the open sea was fresh and clean. Our next destination was a foreign port, Havana, Cuba. This was to be our take-off from the continental United States. We, of course, stayed close in shore again, heading south, using the counter-current inside the Gulf Stream.

Then came the incredible archipelago of the Florida Keys. We saw them as green strips on the sea. I knew they were mangrove forest or coconut palms, although we couldn't see it. They seemed countless, large and small. When we finally got as far west as Key West, we were ninety miles due north of Havana, an overnight trip.

We arrived off Havana before dawn, and had to lay to until light. I was on the bridge with glasses, my eyes glued to this strange foreign land. I saw faint lights which turned out to be fishing boats. With dawn came a mist and there were mountains rising from the shore, not high and old and rounded. Boats were hauling in fishing nets just offshore.

My first foreign country. I was not yet twenty-five years old.

We crept into the harbor. To the east, the soft green mountains, and to the west we could see the glitzy beach hotels which reminded me of Miami Beach. It's a spectacular harbor, many miles long, a very nice landlocked harbor. First, you have to go beneath the once menacing guns of two old forts. The most famous is Morro Castle: prison, dungeon, fort, and lighthouse. Its black, old rock seems to lean toward you; it feels so close, with deep water between a sister fort on the other side. There's an awesome menace in these things. It left an indelible impact on me and it marks the harbor. We could see the sprawling old city, misty in the early day. The harbor itself was a chaotic madness of traffic.

We took on a pilot out near the forts. He spoke no English. All he knew was, "ahead, back, full speed, stop;" other than that he'd point. His immediate command was, "All ahead, full speed." And he pointed at a pier as a target for the helmsman. There were no rules of the road in this place. A naval vessel apparently does not yield space or speed.

I could see immediately that this was going to be a problem. I just quietly took command from the pilot and talked continuously to the helmsman. When the pilot wanted full speed I gave it in numbers of revolutions, which was for half speed. The pilot didn't understand such slow full speed.

It was heart stopping! We had near misses. There were fishing boats coming and going in all directions. There were all kinds of small craft. There were rowboats. There were ferryboats. There were anchored freighters all over the place. There were freighters underway going in and coming out. Sail boats. There were tugs with tows: pushing, alongside, pulling. An incredible mess! We tore up to a pier. I pulled back on the throttles myself. The pilot was spouting again and I told him with gestures that I was taking command. I did not want that character taking this ship into the pier, I was sure he'd ruin us. I ignored his expostulations. So we eased in and got the heaving lines over and there was the usual cautious skipper taking the conn. The pilot shoved a piece of paper under my nose to sign and stomped off as soon as we got the lines on the pier.

The other two YMS skippers (which had joined us) and I got together to pay our respects to the naval attache at the embassy in Havana. He had all the papers and requisitions for us to sign: fuel, water, provisions, all available either at the pier or on order. And we got some advice on what not to do in town, and a warning: "Make sure the crew takes a pocket full of condoms and to wear two of them because the outer one vulcanizes." I recall the exact words and huge grin.

I have a jumble of impressions of Havana from August, 1944. For instance, two other officers and I left the ship in the early afternoon just to go sightseeing. There was nothing for us to do aboard; we selected one officer to handle everything. I knew where we could get a cab. I'd been in this part of town earlier. There were seven cabs lined up along the curb and we went to the first one. Suddenly all hell broke lose. The other cabbies started screaming and flying out of their cars and we had a seven man tag match. It was an incredible slugging match. In the end there was only one guy standing. He smiled a bloody smile, bowed, opened the door and said, "Yes sir, yes sir."

"A ride downtown," we said. We didn't know where to go.

"Oh, no," he said, "See city, see city all over."

"How much?"

"Five dolla', five dolla'."

"Ok, ok."

We didn't believe it, five dollars didn't seem like much, although at that time it was a lot more than it is now of course. We had a heart-stopping ride. He kept going through town the way we had come through the harbor, horn-honking, roaring, ricocheting through the narrow streets. We came to a halt near a street named "Del Prado."

"You see, you see. You come back. I wait."

He was quite a character! He seemed to think this was a good deal. He was being quite kind, we thought.

Del Prado was the obvious showcase of the city. It was a commercial avenue that went for several blocks down from the harbor fort to the old capital building. A beautiful street! It was actually two streets with a wide esplanade in between. The walk was very wide, walk twenty people abreast. The sidewalks were laid in ornamental swirls of tile. In tubs down the middle, there were bushes and trees; bordering flower beds. The trees were cooling. It was beautiful! Del Prado. But it was crowded; in all directions people went, like the harbor; old and young, families, gangs of kids, all talking, laughing, hollering.

This was the first time I heard something that I was to hear on this cruise and in the times I've been in Latin America since, "Cheecklets, mister, cheecklets?" Chicklets, the little candy covered gum, two per package. They sold them for a penny, or whatever was the equivalent.

I remember a five-year old boy, maybe younger. He had a shoe-shine kit. He tugged on the trousers of a dignified man, trying for business. The guy got furious! He turned and he booted this urchin several feet and dumped him on the sidewalk. His little jars and rags and brushes rolled all over. We couldn't believe it! The guy turned on his heel, walked off and the kid sat on the walk. He looked around, nobody paid any attention to him. He just cried silently.

About six teenage girls holding hands (they couldn't have been more than fifteen) charged us and laughingly formed a circle. Then they said, "Pick three for dates, pick three of us for dates." They were in that cute circle around us. It was unbelievable! This was not in our ken; we had never seen anything like this. We stammered our way out of that one. They left with probably some strong remarks about our cojones.

We were buffeted this way for blocks, walked up and down, stunned by this loud talk and laughter. I had never heard anything like this on a city street. Guys would chase girls, young and old, and the on-lookers would cheer. One woman jumped off a street car. She wasn't screaming, she was just looking over her shoulder and running like hell; she was followed by two guys. I don't know whatever happened to them. The city fathers must have gone out first thing in the morning and sprayed the town with aphrodisiacs and other neuro-humors and uppers.

We came back to the cabby and he showed us more of the town, all the fine winding boulevards, the columned government buildings. This was outside the old Havana; Del Prado was just on the edge of the new and the old. There were grand hotels, quite large and elegant; we saw some old ones that were tropical paradises with courtyards and palm trees. The university was a sprawling set of quadrangles. There were apartments and large townhouses. The whole thing was kind of a tired rococo. It was anachronistic; it was not just the town of Conquistadors, but of the monied colonial nobility and commercial barons of the last century that exploited Cuba, a gilded cage. It was steamy and lush and tired looking. We had some peeks into mansions; we could see through gates and into courtyards with palm trees and fountains inside—spectacular!

I made a casual remark to the cabby. "Where do the real people live? What do they do?"

He gave us a curious look. But we swung off the boulevards and suddenly we were in the narrow streets of old Havana again. Clothes lines up above; balconies leaned out and seemed to meet in the middle above us.

There were garbage cans and there were kids and cats and dogs and noise and stink and beggars and drunks. The cab honking scattered the crowds. Cabs played, "chicken" at each blind corner. This was the old sordid, crowded Havana. We screamed to a stop and he nodded across the street as if to say, "Here, you want to see how Havana lives?" We crossed the street. The door opened and a huge black woman filled it.

"Come in gentlemens."

We really did not get it! We were ushered into a room about twenty feet square and lined on all sides with an upholstered bench; the walls were mirrors and so was the ceiling. In the middle was a bed like a prize ring; about a foot off the floor, a tufted, square, huge bed. We laughed and started to leave.

"No, no, no." She stopped us.

"What will you have?" she said. We were in a notorious Cuban circus, known throughout the world.

"What you got?"

"Three girls and a boy."

"How much?"

"Twenty dollar."

We put our caps on.

"No, no, no, two girls and a boy."

"How much?"

"Ten dollar."

We put our caps on again.

"Ok, ok," she said, "One girl, one boy—six dollar. Two dollar a piece, cheap."

So we put down our caps and wondered who the hell the boy would be. She clapped her hands and about a dozen working girls came into the room and sat silently on the benches. They looked tired, dressed in underwear. They smiled, woodenly.

"Which one?" she demanded.

The two officers looked at me and I nudged one of them and he pointed to one sort of blindly and said, "Her."

The selected gal grinned while the others filed out silently; this was not news to them. The boy, my God, who's the boy? But he came in, to our great relief. He was a pimply-faced teenager.

They shucked their clothes and got on this bed and for over half an hour they demonstrated all the positions known to Cubans; it was bizarre! The poor kid was really not making it, and the mirrors absolutely prevented any deception. The gal was a terrible actress, but she tried. She rolled her hips. Then she rolled her eyes. She moaned and it ended in a monumental fake. The kid left. She came over and sat on the lap of the man who chose her. She got a cigarette from him and said, "Smoky in ass." She took a drag and then stuck it in her butt, stood up and exhaled smoke from her mouth, then grinned. She had this enormous satisfaction in her magical climax.

We grabbed our caps and fled, running a gantlet. All the doors in the corridor were open, each with a naked gal smiling and beckoning. The theory is that these guys would be so inflamed with passion they'd pick the first one. As a matter of fact, these guys rushed out onto the street and started taking gulps of fresh air. We looked at our cabby who grinned. Cost us each two bucks.

After that elevating experience with the Cuban circus, our driver said, "Next stop is distillery." We weren't sure what that was but we found out. We went to Mi Bohio. For many years I carried a calling card from that place. We went in there and had an hour. Down we went into a huge cellar, sawdust covered floor, all around us enormous vats. There were maybe half a dozen street cafe tables, chairs and a little bar. The guy rushed up to us and started telling us all about Mi Bohio rum and other hard liquors that they distilled there. We heard everything they had to say. "Now, I want you to taste this ten year old rum." We tasted it. "Now, you taste that. How is it?"

"Good, good." He had a blender of daiquiri, crushed ice in fruit juice and rum. That was a rum daiquiri, a chaser between shots!. He said, "Now you taste this thirty year old rum." It was indeed better, smooth. He told us about Bacardi Rum, that terrible stuff that they ship to the United States. "No, that's not real rum." But of course it turned out that the ten year old and the thirty year old was not that bottle of rum but the mother of rum, the seed from which they made the distilled stuff. They made the best bourbon I'd ever had; I couldn't believe it. Then they had all kinds of liqueurs, banana liqueur—can you imagine banana or pineapple liqueur. These were native fruits. Oh, it was so good! We went through a bunch of these; and staggered out of there. They wanted to sell us a crate, or they could get many crates to the ship; they had their ways. But the Navy absolutely allowed nothing like this on board. I promised, "Someday I'll come back."

In 1948, we did come back. Diana and I and our 2-year old flew down there from Miami; I had my calling card and the cab driver took us to Mi Bohio. This time we took all we could take back with us to the States, four quarts for each person, including the 2-year old. Beautiful stuff! We loved that stuff.

Back to '44. We were shown a good place for supper. It was a family restaurant, dignified but not fancy. The waiters were spectacular and the food was excellent. I remember it most because we had fresh banana and fresh pineapple. No equal in the world. To me pineapple was something that left your teeth feeling like fur. But this just melted in your mouth, including the stalk down the middle of the pineapple which is usually so woody. It was succulent and sweet. The bananas were ripened in the fields, not shipped green to the States. I had coconut ice cream. What a supper!

We wandered around town on the instructions of the cab driver. He went off weary, to meet us later. We found a night club. But an amazing thing happened. Outside on the way to the nightclub, a teenage girl tried to pull me away from my two friends, for her own act. It took a lot to pry her loose. She was an adorable, child/woman in a very winsome way, with big eyes. I got away from her while trying not to insult her. One was continually accosted on the street. That aphrodisiac air again.

Inside the nightclub the liquor was cheap and the floor show was spectacular. The last act of the evening was a solo dance—a sort of seven veils thing. Must have lasted fifteen, twenty minutes. Bathed in blue light and a bolero beat, it was absolutely awe inspiring. A stunning young woman—athletic and graceful—mimed making love to a missing partner. She was awesomely beautiful, on her knees, back bent onto the floor, writhing in ecstacy. She was 10,000 times better than the woman in the Cuban circus that afternoon.

An officer from one of the other ships was dancing with a beautiful woman. She was not your usual pick-up, like the little girl that accosted me outside. She was a grown woman with long blond hair, wearing a low cut backless gold dress and high heels; a graceful, sophisticated woman. We kept motioning him off the dance floor. He finally brought her over; she didn't speak a word of English and never spoke. She just smiled and took our hands very sweetly, very dignified. We were shocked when we found out the next time we saw him; he'd got her for twenty bucks!I think his cab driver did it. It was for dinner, dancing and all night. She was a call girl and spectacular.

We left the night club. My little girl was sitting on the curb next to the cab driver; he'd come back to wait for us. She was still adorable but despondent.

"Please, mister, I have made no business. I am hungry!" I emptied my coin purse—thirty-five cents and pointed to a hot sausage vendor across the street, turned her towards it, pressed the money in her hand and patted her affectionately on the bottom. She ran. Never did my buddies let me forget that I was the last of the big spenders in Havana.

The first lesson on our Latin cruise: there are two kinds of people in Havana, the wealthy and the poor and the wealthy are about five percent. They are very white, very cultured, very gracious and very rich. The rest are either in almost poverty or beyond it. They make countless babies, they worked at it. We'd see a couple go down the street with a long line of children behind them. Astonishing and tragic.

Twenty years later in the 60's, I was on the research vessel *Te Vega,* on the Pacific side of Central America, in El Salvador, in a seaport, La Union. I was sitting there with my scientific colleagues this time, in an open bar off the street. People could walk in and out, no screens, open. A shy, beautiful child/woman sat on my lap. Suddenly, "Buy me a drink, mister. Come with me." I had this incredible recollection of Havana, of that other girl. I was struck by the similarity. Same big eyed, lovely face, fawn body, petite and frail, and that forlorn expression. We talked. She spoke English well.

"Take me back with you."

"What!"

"I would be your slave." She was serious.

"No, don't talk like that." I tried to kid her.

Then in a desperate cry, "Then just take my beautiful little baby girl please, and raise her as your own."

So what has changed? I can't forget that anymore than I can forget that other woman of our YMS cruise—down there in the Latin countries. In fact, I bought a wicker ring from the Salvadoran girl; it's the thing that holds the glass ball above the front closet at home now. It cost about thirty-five cents.

We got out of Cuba with only one case of clap. And we got the filthiest water I've ever seen from the filthiest water barge I've ever seen. We added a lot of chlorine and iodine, so much that the water stank, but I couldn't take the chance. We dropped the pilot off at the dungeon, the last sight of Cuba, Morro Castle. Then the clean and the quiet sea. What a relief! What a noisy, ringing, fragrant, intense impression of Havana! All the senses.

The three YMS's relished the blue waters and the blue sky as we cruised around Cuba through the Yucatan Channel. We skirted Grand Cayman off to the east and the Honduran coast to the west. There was foul ground in between. You have to be very careful where you're going; there are many very extensive shallow areas, shoals.

We threaded our way to the large anchorage of the naval base in Panama at Cristobol Colon. The large anchorages were enclosed by huge breakwaters. There has to be a large anchorage for ships because you have to wait your turn. But we got in alongside a pier. We could sneak into places that the huge ships couldn't.

By now the guy with the clap was very sick. This was before we had any penicillin on the ship, although penicillin shots were known. I asked him why he didn't use a rubber. We told all the men to use a rubber. He said he was from Boston, and a Catholic. I immediately replied, "I know, using a contraceptive is against your religion." I'd heard that one in Boston. "Right," he said. We got him ashore for shots at the base infirmary. He returned with a box of pills. He was ostracized. He had to use the bucket on the fantail rather than the head. The Navy considered that this was time off from duty and he was docked pay for misconduct. That word went in his record: "misconduct."

Other than that I recall three things from Panama. One, this was my first chance to see tropical rain forest. For a guy who'd studied biogeography and ecology, this was a beautiful thing to see. Secondly, the Canal passage itself was a spectacular experience. And once again, Latin poverty hit us.

We had to wait our turn to go through the canal, so another officer and I took the train across to Panama City and back that same night. The building of the railroad must have been back-breaking. It went across long causeways in the water and bridges and cuts through the mountains, switchbacks. The train was pulled by a steam engine and there were a dozen rickety cars with the passengers in some and flat cars that carried crates.

Through the open windows, the soot and the embers and the noise and the smells came in. It was slow. It creaked and crackled and pounded on the straightaways. It was amazing what the cars would carry. We had local passengers, but there were military people like ourselves and locals with wild luggage. They'd have a crate of produce or chickens.

All around us the jungle was a solid wall of green. You could still see the terraces where they had dug. Two or three small villages—awful places, hot and steamy. But it was text book forest and I saw these huge tropical trees with the flat tops and the other trees gathering underneath in this classical tropical rain forest stratification. That excited me! It struck me then that the local folk must have seen the same forest as so ordinary; always like that, everywhere.

It's not hard, looking out, to reconstruct the tragic difficulty of building that canal. You can see that the earth moving must have been unbelievable. It's hard to imagine doing it today much less at the turn of the century!

Even worse than the hardships of construction was disease; many thousands died of dreaded and fatal illnesses of unknown causes. As a youngster I had read De Kruif's *Microbe Hunters*, and there is a Gorgas hospital in Balboa as I recall. How high the stakes and how dazzling the science of tracking down insect vectored diseases, right here in Panama, where I was riding a train. That poor country, where U.S. muscle and brains tamed the hostile environment, will in your lifetime be devastated by renewed environmental hostility. Deforestation, silt.

We reached Panama City. A shocking dichotomy was obvious. Balboa was off on one side. This was the Canal Zone. And the other side was Panama City. The Canal Zone looked like a southern California town, Panama City like the old part of Havana. Two different cities—squalor vs. loveliness, crowded vs. empty. Balboa had mowed lawns, large buildings and housing for the military, for the Canal pilots and so forth; boulevards wound around; we saw flowering bushes, quiet family groups, individuals sitting on the grass.

Over in Panama City: noisy, crowded slums compared to Balboa, dark skins compared to the white skins. Panama City was like old Havana, without the charm and energy. The cathedrals weren't grand. They were big and the interiors were stunning but the outside smelled of rot and the mortar seemed to be crumbling and there were stains of fungus and algae. It was dreary and the people seemed more surly. They certainly weren't the lively people we saw on Del Prado. I didn't see the whole town, but the parts of old Panama City we saw were very crowded and mean. We went to visit a night club but the night club was nothing like the Havana night club. It was sleazy and noisy. There just wasn't a joy of life. There wasn't the class. One thing I did here that I didn't do in Havana, I got a bottle of Canadian Club Whiskey. No taxes, duty free, cheap and I put it in a box labelled, "Not to open till VJ Day." That would be a historic moment and I was looking forward to it. Actually, we could not wait our turn to make the passage through the Canal. Panama City saddened us.

3. The tropical Cruise Ship

We YMS's formally joined the Pacific Fleet at Coco Solo, Panama Canal zone, on the 1st of September, 1944. We'd had a conference and got the word on how we were to make the passage, what the procedure would be, and we got orders to proceed without a stop at Balboa, right out to sea and on to San Diego Navy Yard, reporting to the Commandant there. There was refueling scheduled at Corinto, Nicaragua and Manzanillo, Mexico a third and two thirds of the way up the coast to San Diego.

The passage of the canal was a barrage of sights, some surprising. Actually, the locks are few. Most of the Panama Canal is a fresh water lake, the fifty or so miles of Gatun Lake. The locks are at either end: the Pacific and the Atlantic end. The ship was raised up eighty-five feet (or something like that, I don't remember the exact number) to Gatun Lake. Then we went across to the Pacific side and were lowered to the Pacific. The astonishing engineering feat of these locks is that they don't have pumps. The lake, being eighty-five feet higher than the oceans, is its own water pressure and it's the water pressure that runs the locks. So its locks are gravity powered. Actually, this Atlantic to Pacific Canal is southeasterly, not westerly. It's a funny thing the way Panama bends there to east and west so that you go north to south from the Atlantic to the Pacific.

In the locks, the 3 YMS's were lashed together with mooring lines and only the middle of the three had power, and that's where the pilot was. So he took us into the locks and took us out up on Gatun Lake. Since we were on the outside we were just spectators. We looked after the lines and we manned the fenders if we got too close to the walls of the locks. We slowly rose up to Gatun Lake in these enormous locks. You know, you could put a battleship in the damn things. They're over a hundred feet wide and many hundreds of feet long. We went with a whole bunch of other ships, one or two fairly large ships side-by-side and then a whole bunch of small ships. They filled the locks. This took a lot of doing. It was noisy and busy. It was thrilling. The locks gushed with rising water, misty and dank.

We uncoupled up in Gatun Lake and went on our own steam in a column, down a long line of heavily buoyed channels. You couldn't go wrong, but we were in a vast open lake. The shores were distant and there were islands, former mountains but when it flooded they became islands in the lake. Almost no human existence. And you could see tropical forest all around. I was looking for monkeys with the field glasses and quite often we'd be close enough to see waterfalls gushing out of rocky mountainsides.

One of the real big thrills to me as an ecologist was to see Barro Colorado Island; the Smithsonian Institute has a field biological station there. The channel went half way around it. I had the glasses on it all the way. This was a shrine, an ecological landmark. I'd read about Barro Colorado Island. One of my professors had studied there and written on temperature, humidity and wind.

The Pacific side of the canal is steep-sided, a dredged, narrow channel which ends in the Miguel and Miraflores Locks. Here you can see the scars of digging very clearly. We nested again, went through the locks, sank imperceptively down as they drained the water. We dropped the pilot off at Balboa. It took most of day. The pilots were very professional, low key, intense, pleasant. They handled everything. They'd be on a big battle ship one day, then a huge tanker, and then 3 YMS's bundled together.

At last, we turned south into the Gulf of Panama, went by the Perlas Islands, then west, then northwest to San Diego. And we were in the sparkling blue water and under the sky of the Pacific and our Latin vacation continued.

In our trip from Panama north to San Diego, we had three thousand miles or so and it took us a couple of weeks.

The west coast of Central America is spectacular from the sea. The mountainous spine is so apparent that even when you can't see the shore, you can see peak after peak after peak rising in the distance. These could be ten to fifteen thousand feet tall. It's part of the enormous spine of mountains that runs through our hemisphere from almost pole to pole. We navigated by mountains. I remember Momotumbo which is so sublime, beautiful, and perfectly pointed—a still active volcano. I have since looked in close on some of these places like El Salvador. They smoke, they're active, they have fumaroles, cracks that hiss out a smell like rotten eggs, sulfides.

I had a birthday. Nobody knew it; I certainly didn't tell anybody. All of a sudden there was an all-hands party with a cake. Only the yeoman, who had the records, could have known my birthday. I was touched. People were calling me Captain now. The original crew, "plank owners," no longer saw their former Captain as the real Captain. I was no longer "Mr. Bovbjerg, sir." I was "Captain."

We drilled and drilled. Seemed more urgent now that we were on the Pacific side. No enemy subs; they had never really ever been there. We "put out fires" in the paint locker and in the magazine. We used wooden shells for the 3-inch gun, beautifully turned maple shells. The hot shell man wore big special asbestos gloves and tried to catch them as they popped out of the breach. We fired the twenty millimeters and .50 calibers but it was hard to shoot at something. We would throw boxes over the side and shoot. I shot my .45 pistol at such boxes—never even came close. Those are ridiculous guns. They were designed, I'm told, for Moro warriors who had run "amuck," but from no more than ten feet. They were so powerful, that they stopped a man who was swinging a bolo knife.

We had a great idea that kept the guys happy. We challenged them to remember how to make and fly a big kite. That really tickled them—everyone remembered how much tail it would take, how much bow. We had wood; we had coat hangers; we had paper; we made them up. They painted Tojo's face on these things, and flew them from the fantail with several hundred feet of light line. The kites always had the wind of steaming at twelve knots plus a brisk, steady breeze. So these things danced and dove and swooped and climbed, and then all the machine guns would let loose. It was a great challenge because they never knew when it was going to duck to the right or left, dive or zoom up. We did shoot them down, but not instantly. There emerged a certain respect for moving targets. They did learn to lead; they could sense when a dive was coming and lead it. They learned how to hose with tracer bullets.

But among the times in that three thousand miles there were two outstanding memories and those were the stops in Nicaragua and in Mexico, not tourist towns, no casinos, unlike Havana.

Momotumbo was our beacon to Corinto, Nicaragua. We could see it from forty miles away. Then on into the strange harbor in Corinto, full of turns, and narrow channels and buoys popping up that were not very well marked. Very snug harbor, remarkable. Many bars and islands on the way in but then we only drew one fathom. We did finally come to the right pier, the three ships nesting together in the customary way. It was nice to be ashore again. The pier was old and the planks were loose. So guys going ashore with sea legs that were kind of rubbery also had to watch their step on an old pier.

The U.S. Navy had a sea plane base up the harbor, up in the more shallow water. We had radioed them for instructions and told them when we'd be arriving. One of their men came down that morning and arranged for immediate fuel and water on the pier; it didn't take long at all. We would spend the night.

The town sat on the bay, and there were hills to seaward, so we were not only snug in terms of water but also in terms of onshore wind. It was a sleepy and decrepit place, though. There were no customs officers; there was none of the big-citiness of the other places we'd been, and no taxis.

There was an amazing sight at the base of the pier when we first walked out. About fifty guys were squatting on the road. Each one hammered a rock that he could hold in one hand and the chips were flying. They arranged themselves in a long row and the chips from the rocks were resurfacing the road. The guys had straw hats, rags for clothes, sandals; they smoked; they laughed a lot and you could see the results of the work; the road was indeed getting new crushed rock. Slowly.

The town was sleepy, squalid, dirty and the people did not have the frantic pace of the Cubans, nor the surliness of the Panamanians. Families sauntered around. Women came by with huge loads on their erect heads, carrying them so gracefully. There was laughter; they chatted with each other. Ox carts were the major movers, an occasional car. Dirt streets had deep ruts, puddles, manure. Burros were laden and ridden. The buildings were stained with old green algae; I'd learned to look for the sort of green stain on that old yellow stone. There were no narrow city streets and crowds; it was a large village, a port town with warehouses and ships being loaded and unloaded.

I went to an apothecary. I had run out of nose drops; had some sinus trouble in those days and there was new stuff called Paradrine. Then I also needed anti-diarrhea stuff (I was always worried about that). The drug of choice was Paregoric which was, of course, dope. You couldn't get it in the States. A teenage daughter of the druggist smiled. She spoke very precise English and she knew the drugs; they had Paradrine and Paregoric.

The vendors descended on the ship. This was not the usual day. The word had spread that three U.S. warships were coming in the channel and we brought everyone out who had something to sell. Instant bazaar!

But the big event followed the captains' morning visit up the bay with the commanding officer of the seaplane base. We paid our respects and asked about liberty for the boys. "That's all taken care of," he said. "El

Phenix." He told us that the doctor would be making his daily inspection and that we, the skippers, could attend if we did it officially, in other words wearing sidearms and full uniform.

El Phenix was not hard to find; there was a sign over the door. It looked like a warehouse or a walled factory. The wall was high and covered with vines. Massive doors were very prison-like. Inside was a courtyard lined with palm trees; a fountain bubbled in the middle of a flagstone patio. All the walls had cabanas leaning against them, opening up out onto the courtyard—a whole bunch of little single rooms with a door and a couple of windows. One end of the walled enclosure was roofed, but open to the patio. They had a bar, tables, a dance floor covered with sawdust, red-checkered tablecloths—very neat. We were expected and the girls had started sweeping immediately. The place was beautiful! There was a nice breeze blowing through it; the palm trees were rattling.

When anybody went into El Phenix, they were greeted by a U.S. Navy Corpsman. He handed out condoms. He took the I.D. card and followed with a "short arm" inspection. That was old drill to me; all the time I had given physical exams in Boston: "Skin it back, Mac, milk it out." On leaving, everyone got a venereal treatment before getting the I.D. card back. We were official visitors so we didn't go through that.

The doctor hollered when we got in and all the girls emerged, peeled off their clothes. This was not a Cuban circus, this was a physical exam. They jumped up into the fountain and splashed about as they bathed. There were screams and laughter. Then they lined up for the doctor's exam. He could put anyone on a sick list and that took them out of circulation, until he said they were OK. They had cut VD to just about zero.

The girls were young—some of them were very young—and there were a couple in their thirties. Most were local. According to the doctor, a lot of the local girls were making enough money for a dowry. There was money to be made here, cheap as they were, to get a dowry in a year or two. Some of them were paying off father's debts, which gave rise to the lousy pun that it really was a "penal" institution. Most of them were fairly sturdy, brown, peasant stock. Big smiles all around. Some were from the U.S., end of the line call girls, or one jump ahead of the law. There were some dyed red-heads and blondes. They were sort of sad. The local girls were far more giggly and happy.

By early afternoon I was back on the ship with the news. I told the senior men, described the place, the rules and the directions: one block this way, two that way. I also warned them to tell the men to stay away from the local women. I just could not see the druggist's daughter being accosted by our guys after a few beers.

We made one rule. After all, the United States was in a state of war; we couldn't just desert the ship, so half the men had to stay on board. We could get underway; we could man the guns. We had to have some security. We, for instance, ruled that no one would be allowed on deck who was not in the crew. So we had an armed deck watch, and an armed officer. But I said (believe me, most innocently), "It doesn't matter if a whole watch goes on or off; if you come back alone then your running mate on the other watch can go, anytime, as long as half the men are on the ship." It was a crazy mistake.

So half the crew raced down the pier, and the watch on duty sulked, but figured their turn would come.

I walked around town and took some pictures. Folks wanted to show off their English. This was small town stuff and I liked it. But the poverty hit me hard. In the tropics you can survive on fruit and fish, which are free. Housing can be a piece of tin on a crate. There is no cold winter, but there are amazing diseases. Everywhere people looked like they had skin sores and scabs. Many were crippled. Their bodies were parasite hosts. Sanitation was very poor. They could just survive with little income, but people seemed nice and I enjoyed them.

I got back to the ship late in the afternoon. Bedlam and drunken hilarity! The men straggled back, frequently on the arms of their new girl friends, so that their shipmates on the other watch could sprint down the pier. Half the men were aboard, which was all I asked. The saying was: "Screwed and no longer in the mood." They passed out on deck, and the officers were busy lining them up with life jackets for pillows. After all, we could hose off the puke tomorrow.

They had been sold string bags of coconuts and bunches of bananas that must have weighed fifty pounds. Parakeets, fish, baskets, straw hats, sandals, and "Cheeklets, mister." All four officers were out on the deck. It took the four of us to keep order, until they'd passed out. And then there was this long row of corpses in the sunshine, sweating and puking and sleeping with a smile. Then someone would return, and he'd wake his mate, and change places on the deck, and off would go the next guy, the liberty hound, staggering back to El Phenix after about three hours rest. The ship was in bedlam, but also a Mid-Summer's Night Dream.

How very different this was compared to the Cuban house. This was a rollicking party compared to that sordid and sad affair on the back streets of Havana, or compared to the lonely sweet waif peddling on those streets.

The book, the play and the movie, *Mr. Roberts* captured this event in one great scene, when they lowered the returning watch in a cargo net, and a couple of guys came down the pier with a goat and another one rode a motorcycle off the end; that is just what went on. Pandemonium!

The whole town was watching, knew what was going on, and thought it was hilarious. By midnight there was hardly a sober crew member on the ship, just the four officers. The bosun was a pretty sturdy kind, and I got him to find, in his locker somewhere, armbands that said, "Shore Patrol," and billy clubs, which we had but had never used. I put on an arm band, and got a billy club, and he did the same; we rounded up the last of the liberty hounds. The bosun was loose, but in command. We cleaned out El Phenix. You can't imagine the wailing and pleading of the last guys on the dance floor. By 0200 all were aboard. Four solemn officers, thirty passed out crew.

In the morning the four officers had to get the ship underway. One on the deck handling the lines, one in the engine room, one on the helm, and I was on the side of the bridge. Some of the crew made it up to their stations. Others felt the engines start and the ship beginning to rock a little bit, so they came up, duty bound, staggering to their posts.

We retraced the tortuous track of the way in, and we got to the open sea. Oh, what a relief! How good to be at sea! Clear air, so important, and the water was spanking. We hosed the deck down and scrubbed it. Clothes were hung out to dry. Some of the crew trailed clothes in the wake on long lines. That's a tried and true Navy washing machine.

What to do with all the parakeets? Incredible, the number of parakeets we had aboard there. "Well, we have to shoot them all," I said. We could not take them into the States, because parrots are famous for carrying psittacosis, a bird disease which people can get. I said they'd never let us in. The men didn't want the birds; they didn't even know to whom they belonged. There they were, all up in the rigging, on the radar, and crapping all over the deck. So there was "pop, pop, pop," target practice. No one mourned those birds.

We ate the fruit. But I insisted that the cook only use fruit that we could peel. Sandals and straw hats went into lockers.

There was only one thing left I suppose; a lifetime of memories was left. Every man, from teenager to family man. They were bragging and talking and laughing, and they told tales later, in the dark of night. How the red-head fought for them, how the cute little Indian girl was going to commit suicide if he didn't jump ship for her. How they danced and fed the juke box. How many times? They played in the fountain and under the moon. Palm trees, a fountain and a willing woman. I heard those stories a year later. In the very hushed dark, in some corner of the ship, there'd be two guys talking, "Remember the night at El Phenix!" I am sure the memory still stirs.

Landmarks for Manzanillo, Mexico were two volcanic peaks. That's all we could see on the eastern horizon, Colima and Safa. This was our last provisioning stop before San Diego. Nicaragua was now days behind us. We turned in, and as the shore rose up out of the sea it turned out to be desert, though we were in a rain storm as we approached the shore. We saw cactus, and the vegetation was what you might graciously call, "chaparral." There were huge mesas and black cones all about. The town sat on this coast behind a large beach and this was black sand—at least very dark anyway. We were going to see more black sand and volcanic islands but this one was the first. It was a thrill.

As a harbor, this was not snug. Nothing compared to Corinto, Nicaragua. This was an open bay. A huge break-water protected the piers and anchorages. However, it was difficult to anchor and maneuver because there were strong tides and frequently a heavy sea. We were instructed by the port authorities in town to anchor some distance off the pier and then back down to the pier while letting out chain. Then we could get lines over the stern, warp into the pier, and still have chain out to an anchor. When we needed to move, all we had to do was cast off and we'd swing out on the anchor away from the pier.

We were expected, and the fuel and water started right away. But there was no El Phenix. On the other hand, somehow, the men did not mind too much. Some of them wanted to go to shore, a lot of them didn't. Those that went ashore went for a beer. They had bars and the mexican beer is really good. But they really didn't seem to crave another night on the town like they had in the unforgettable El Phenix.

A couple of the other officers and I walked around downtown. It was a mess of mud; there were few sidewalks. How dirty and dusty it must have been between rains. Here again there was poverty. This was a caricature of the sleepy Mexican village. The guy sitting with a huge hat. It was a city of thousands but it didn't seem like it. There was no big business district, although there were warehouses. It was a port city and had a railroad that went to Guadalajara; it had gotten washed out from that flood with the rain.

The smell of the place was something awful. A refuse wagon came by, pulled by two burros with a man sitting on a high seat. And then I noticed a very delicate looking lady with one of those big lace scarves, a *mantilla* over her head. This classical Mexican upper-class woman was walking and she pulled her skirts up above her ankles. Then out came the handkerchief and she covered her face with it. I said, "What is she doing?" Well, when this honey wagon got by us, we knew what she was doing and we got our handkerchiefs out awfully fast too. Another woman passed, an elderly woman, and all of a sudden she squatted in the street. When she got up; she left a wet spot. Dogs were yapping and crapping. Burros, oxen all leaving manure; trash and garbage everywhere.

We barely got to the larger *haciendas* in the surrounding hills. They were, of course, very much nicer, belonging to that upper five percent.

Back down town we found a hotel. It was old but had a bar and it had a cantina that opened on the corner to two streets. And the breeze came in an open side, no screens, it smelled; flies buzzed. There were occasional street noises, almost no cars. We drank beer. By now we could struggle with the words: "cervesa, por favor, amigo." "Uno mass por favor." They smiled.

Then we heard music far down the street, but we didn't pay any attention. It got louder. Word must have gotten out about the gringos at the hotel. A mariachi band: trumpet, couple of guitars, percussion: sticks, gourds—stunning music! They stood in the corner and they played and then they would sing. They broke our hearts. I have never been so moved by music. I have never been able to find that kind of music on a record. I wish that I could. The pace was slow. The harmony was excruciatingly close. I don't know music but they seemed to structure it. They had moods and climax with that haunting harmony; they moved slowly together. We kept putting dollar bills in the hat. They kept playing, smiling, with sad faces that said, "This is it, this is life."

The beer was cold and very tasty. At least it was sterile; we wouldn't have dreamed of drinking anything else. Fans turned silently above the place. It had a tired elegance, had been lovely perhaps at one time. We brushed the flies away. We didn't talk, we couldn't. We sort of cried inside. Why? These poor people! Such poor people with such rich music! They bowed and went out the door. They and their music disappeared down the street and we returned to the ship. We did not talk. I cannot erase a single element of that memory and my eyes are wet even remembering it.

Course West, Speed Ten Knots

PILOT HOUSE AND "FLYING-BRIDGE"

4. Course West, Speed Ten Knots

Our all expenses Latin American cruise was over when finally we steamed into San Diego harbor in mid-September. Still not even a month. Spectacular landlocked harbor again. Point Loma and North Island form a perfect channel which bends to stop any stormy seas from the west. We tied up in a nest again on a huge buoy in the middle of the bay which could have held a battleship. Paid our respects to the Commandant of the district, Admiral, very proud, very cordial. "Let us know what you need." Very nice guy.

The striking shock was being back in the States after weeks on our Latin cruise. We were shocked at the paved streets, at the sidewalks, at the lights, at the stores with plate-glass windows instead of bars and shutters. So many cars, even with the rationing. But mostly the shock was seeing so many middle-class people who seemed happy and well dressed. We didn't see the poor people, of course they were there, but that's not what constituted most of what we saw. That struck me especially when I took a day off to visit my very best boyhood friend who lived in suburban L.A. He had a wife and a small child, a couple of years old. I revelled in an easy chair on a carpet and a dinner with a family and a yard and an orange tree. It seemed so long since I'd had even a touch of domesticity in my life. It was only a month but how I missed my wife. She could not come to visit because we could not say where we would be or when. "Loose lips sink ships"—we both understood that.

This domestic jolt was submerged by the day-to-day Navy Yard business. We knew the noise and the scramble and the chaos, the in and out of drydock and all the repairs by the naval technicians, all over. Our supplies were all aboard and the ship was a shambles. We got two new twin .50 caliber machine guns on the fantail. All kinds of dirt and shavings and wire and people lent a certain frantic air. We were headed now for the Pacific War, maybe for years. Last chance for repairs.

Last chance at gaiety and buzzing around town, too. Drinking was the thing to do. None of us had cars so it didn't matter whether we were too drunk to drive. We'd get a cab back to the Navy Yard. We especially liked the top of the Coronado hotel on the hill. What a view over the harbor and the Pacific, especially at sunset.

We agreed to meet one night in LaJolla to try a very famous restaurant we'd heard about. I went to the Grant Hotel downtown where all the buses came in. I waited and I waited. Then I finally asked a guy, "Where does the LaJolla bus stop?" He said, "Hey, there's one just now pulling in." How did I know that you spelled "LaJolla" for "La-hoya?" Great Spanish scholar! Finally got out there but by taking the bus to the end of the line and hitching a ride with a whiskey salesman. I don't think I'll forget him. I don't know how we got back, cab I guess. All this night life, so shallow but an imperative.

Our last night we went to a Mexican night club, the hottest one in town. The other officers had been there. I had not. They put me in a chair right next to the dance floor with the best view. When the climax came of the fandango I was grabbed by the prima ballerina and led out on the floor to dance with her. My friends were doubled over. I never was much of a dancer.

Our day finally came for briefing with the staff of the convoy we were going to join for Pearl Harbor. We were going to be escorts. They instructed us in our duties, and the schedule of the days were laid out and the convoy regulations were spelled out. Ran into another escort skipper; a guy who was on the university swimming team with me. We were so young, God! But this was now serious stuff. Our duties as escort ships were anti-submarine and anti-aircraft and that was not kidding.

Of course the last of our big ports of call on this long trip to combat was Hawaii. The fabulous island of Oahu. We arrived after a smooth passage, and only had one incident I want to tell you about.

On the fifth day we had to refuel at sea. We were running low; so were all the other escorts. YMS's were never intended for this sort of duty on the high seas, but there we were. We were to get our fuel from an LST, landing ship. I had read the Bureau of Ships manual on how to refuel at sea. But I had not even seen the operation, much less done it. And it's one of those things that takes a lot of seamanship, not best done for the first time never having seen it, I can assure you.

The manual underlined that commanding officers take the conn, the control, give all orders, including the orders to the helm. The tradition had always been in the Navy that you let the helmsman keep station on the oiler. However, now they said that practice should be abandoned. The captain on the wing of the bridge was a better judge of station keeping and should give rapid orders to the helm, even to half degrees of heading. Made sense to me. I explained this to our number one helmsman, who would be at the wheel. He agreed. (He was Regular Navy and very good.)

The exercise is a tough one. The helmsman and others of the bridge gang, the engineer and his black gang and the bosun and his deck gang, must coordinate. The LST, our donor, was to keep a course and speed; they blinked over to us what their course and speed were. But it's tough in a heavy sea; we were both rolling. Trouble is, they were more round bottomed than we were, I think.

We would put over a spring line; the bosun and I agreed. He did not think that lines fore and aft would hold in that sea. A spring line is a diagonal line, and it would restrain us but not snap. The angle would tend to pull us in to the LST and also not let us go out very far. The deck gang manned the line for tension, to keep it on a steady strain. The burden was put on the skipper and the helm. We had to stay very close, but not collide. That was not easy—understatement!

The black gang had a hose and nozzle run out. It was hung on a line, a restraining line, so that the hose could have a slack, but not sink. That line was handled by the black gang to keep just the right strain on the hose.

We pulled in alongside; the heaving line went over, then the heavy line. I went into that trance and played with the orders to the throttles and the helm till we got a sustained position and held it quite well. Then the hose snaked over on the lines, and we were connected. They started pumping fuel into us. The spring line restrained our pulling out, and I pulled closer. But my intuition rebelled at that rolling embrace of those two ships. We were steaming slowly, but the waves crashed between us as we compressed the sea into that seething resentment. Then we would suddenly yaw out, and then seemingly be sucked back into the LST. Then we rolled away. I began to find a rhythm, and there was a certain rhythm; the sea has a rhythm. I kept up a chatter with the men on the helm and throttle. They repeated the commands with an "aye," very carefully. No light touch here. I gave half degree headings, as BuShips had advised. The deck gang watched the spring line; the windlass was turning, ready to ease or strain, take up slack. We made sickening lunges and rushes at the LST; the whole ship seemed pushed sideways at times. It was a long, long, exercise.

The loop in the fuel line would occasionally snap tight as we both rolled in opposite directions. The black gang would let the slack come out in the line, and then take it back in. Nice coordination. It was a non-stop nightmare. The minutes crept by till time did not exist. The bridge was a humming machine. I never took my eyes from the spring line, from the hose, or those two hulls. I did not look at the sea or the sky. I threw my voice over my shoulder. They answered me; their replies never wavered. Then finally the chief engineer hollered up, "Topped off, topped off, skipper."

I hollered back, "Secure." The hose plopped over, limp, and the line took hold. It was pulled aboard. "Let go spring line. Left two degrees rudder. Steady as she goes." We pulled away slowly. "All engines ahead one half." I still had not moved from the wing of the bridge.

The bridge gang on the LST waved. We waved back. Then I turned and went in to chat with the bridge gang. They were relieved; they were limp. We all were. Back on station I gave the conn to the officer of the watch, who'd been down on the fantail. I asked him to blink over to the LST, "Thanks," which he did.

I started down the ladder and overheard one of the chiefs complaining and laughing with a friend. He was ridiculing a captain giving half degree course changes to the helmsman. Not the way a Regular Navy skipper would do it. I just couldn't pass that by. They looked at their feet as I walked by; they knew I'd heard. So I asked the chief to come in for a visit in my stateroom, showed him the BuShips instructions. "Did you ever read those?"

"No, Sir."

Then I told him what we heard on the bridge after we had secured. That we were the last escort to fuel, and that we were the only one that did not break a hose or collide, and the only ship to completely refuel. We agreed that I would endorse a request for his reassignment in Pearl Harbor. Nothing escapes on a small ship. That word got around.

In Pearl Harbor we got replacements for our most senior men. And these were "plank owners;" they felt they knew the ship before anyone else, but the ship really did not need a lot of senior guys now; we had had our breaking in together. And someone in BuPers or in Pearl saw that and took the necessary action. The new younger men fit in well. It allowed for more upward mobility and responsibility. This was good. All this was part of the maturity of a crew and the captain, now for five months. They were ready and Hawaii was the last port of call with a Navy Yard to ready the ship.

The islands were so welcome, so beautiful from the sea. We'd sailed by Molokai, the leper island, with such a rolling green forest. Diamond Head loomed up so prominently on Oahu, and there was the beach and the surf at Waikiki. The whole crew came out on deck. The escorts were the last of the convoy to enter Pearl Harbor. Entering that harbor was our first real coral reef sensation. It's more than a sight; you can sense it all over. The reef is colored; the pale, clear water; and the live reef and the white coral sand; and the blue and the turquoise and the dark shadows and the breaking sea. And that's what marks the menace to ships. But that also marked the excitement of a biologist.

Another spectacular harbor. Such a narrow opening, then the water opens in all directions. I'd studied the charts, and I had reviewed the Jap attack, the damage. But three years after the attack there were almost no reminders.

Overall impression of this staggering base of the Pacific Fleet: the endless piers and the warehouses, the barracks, the offices, the dry docks. Huge cranes! The place was crawling with trucks and jeeps. All of this massive activity, this mass of warships! Cradled in this gentle bowl of mountains, deep green pineapple fields, and bright red patches of tilled earth. I have not seen it like that very many places. An overwhelming first impression was that of our own insignificance.

We were directed to our berth, another nest of YMS's. We were a grain of sand on the beach, one tiny effort in this massive, uncaring war. Even though we were only a few days from the San Diego Navy Yard, the Pacific Fleet Command invited our requests. The ship and gear were inspected. But the thrust was on personnel. Our gunners went to school. I went to school.

Magnificent twenty millimeter cannon simulation. Both gunners and officers practiced. It was a full size twenty millimeter gun in a domed screen, projecting movies of Japanese planes attacking. And they would come from any direction, with sound. The gunner would swing, and learn to lead, fire and stop. The gunfire was very realistic with gunfire racket and tracers on the domed screen. It beeped if the plane image was hit. Planes came even from behind. You'd hear the sound behind you, and you'd swing the gun around, and there the thing would fly right over your head! You almost always missed it, could not get in front of it. Great simulation!

We also had a spectacular anti-submarine simulation for the bridge crew and officers. The skipper went too; we went as bridge teams. The simulation compartment was a mock-up of a bridge; there was a helm and compass; there were throttles, and sonar gear. And next door, we couldn't see, the instructor had a plot sheet, electronic dots on it. It was a map with our ship and a submarine plotted. It recorded our course and speed, and therefore kept a log, a track on this plot. They maneuvered the sub to escape; we tried to keep sonar on it and attack. All of our activities were recorded.

Our sonar man would hear: "sproing-oing-oing-oing, pip," and then he would call out, "Contact twelve thousand yards, bearing forty-five degrees relative." I would call for right standard rudder, and then he would say, "Dead ahead, target, ten thousand yards." The hunt was on. And it was very realistic. A half-hour later, after we'd chased it around, we finally got a good fix on it. We fired a pattern of depth charges, which were recorded in the next room as an attack. The game was over and we went next door and the instructor would show us the tracks of the submarine and our ship. How badly we missed! He softened the humiliation with very good instruction. We did this, team after team.

I also got a lot of flash cards of Jap planes. We were getting the guys ready, and ourselves.

The ship was ready.

The impact of Pearl Harbor was the fury of activity typical of all huge Navy Yards. This was the naval upkeep of the Fleet; resupply, repair combat damage, new construction. We were always overwhelmed when we got into these places. After many days I could be still lost in any one of these big yards. Hundreds of buildings; railroad tracks going everywhere; lines of railroad cars unloading, freight cars. And huge factories, train tracks going through them; they could be repairing battleship guns. Whole sections of ships came in. They were fabricating. Machines shops. Noisy boiler factories. Racket! Riveting. Welding; sizzling blue light. There were laundries, mess halls and barracks and guard posts for Marines. And Ships' Services. Golf courses. Recreation places, officers' clubs, enlisted clubs, movies, bowling alleys. There was always a Post Office somewhere; everyone wanted to know where that was. There were operations offices, where the brass were sitting around planning. There were brigs, jails. There was always a Marine base next to the brig. And hospitals. A BOQ, Bachelor Officers' Quarters, places that we sometimes stayed when we couldn't be on the ship. There was a motor pool for cars, trucks. Boat pool for small boats to get around the harbor. And out on the water there were tankers and tugs and water barges.

And then there were sanitary facilities and power plants and telephone centers and there was intelligence and communication centers. The place was a maze of poles and wires. Fire house, shore patrol, port director, signal tower. There were dry docks. Office buildings. A thousand typewriters banging. There were banks. My God, I could go on forever! These were cities.

These were cities, but no through streets. Streets had heavy traffic. Huge cranes on stilted legs whose wheels ran on wide tracks down the streets. They were so big that a truck could go underneath them. They would go very slowly, carrying heavy stuff from factory to ship. Gun turrets, a new bow, boilers; a bell clanged whenever they moved, "Get out of the way." And trucks were always roaring by and jeeps were rumbling by and people were going somewhere, they knew where. I did not know! Here was a mystery to most strangers, lost and intimidated by bedlam.

They were dangerous places. I once watched a welder in a dry dock get fried. He shorted out with his electric welder and turned beet red; he sort of exploded, then became just an ash. I saw a huge mobile crane carrying a gun too heavy for it. It tipped and this guy got trapped underneath it as it went down in a crash. A big puddle of oil and gasoline spread out; blood and oil did not mix; not a pretty sight. These were tough places. This was heavy industry. In a hurry.

The most drastic event for ships was dry docking. We were too small to bother the big ones. Alternatives to dry docking were marine railways, a set of tracks sloping down into the water. The 353 would ease up onto a wheeled platform; they'd brace us up and then pull us up the track with a big winch. We'd be out of water and all encased in scaffolding, big timbers.

The floating dry dock served us several times, and that was just a huge floating bathtub which could sink or bob up depending on the water in the tanks, water pumped in or out. The gate closed behind us and we were inside. Water was pumped out and we were dry. The ship was propped up with timbers, took careful handling. Of course the crew of the dry dock was very good.

Suddenly we were in there, in this strange, alien environment of air instead of water. Everybody turned to with scrapers and paint, but we worked on everything. I was looking for ship worms, actually a small clam, the bane of wooden ships. The engineers would attack the sonar head, the shaft, screws and rudders. We replaced zinc plates back there to protect the bronze screws from galvanic corrosion. The carpenters plugged the ship where worm damage was bad. Then the bottom was dried and painted; anti-fouling paint. We got out as fast as possible; another ship had to go in. Then the dry dock would fill or we would slowly go down the track if on the marine railway. We could feel that first flotation and a little bit of rocking. Suddenly you felt right again, being afloat.

Meanwhile everything's crazy. You can't cook, you can't use the head; there's no shower. You have to walk a plank out to the ship or climb a tall rickety ladder made of two-by-fours. Don't come aboard drunk when you're in dry dock.

The same craziness goes on all over the ship in a Navy yard. Everything's overhauled. Somebody knew we were coming and had all these people ready for us. It was amazing! Specialists came, the electricians and steam fitters. And the crew was working hard chipping and painting and marlin spike work was done on the lines and the blocks, the davits. Everything at once. It's just chaos! And I recall the saying, "Man, it was such a mess I didn't know what to do, shit or draw small stores." Or, "Copasetic, Mac," meaning, "Cool, man." Or "Hubba, hubba, fella." Answer, "Goodyear rubba, fella." Like every era, slang prospered. The workers sometimes would make these little drawings of eyes and nose above a line, and that would say, "Kilroy was here." These people descended on us like rodents, ravished us and then left, leaving chaos behind, a litter of crates, wires and cartons.

Outstanding memories, not of Pearl Harbor, or the Navy: I remember visiting the university. Oh, how I longed to be at that university. I was still a "student," three years later. What a beautiful campus. Actually in 1941 I had applied for grad school here. I wanted to get the hell out of Chicago, badly. But the draft board nixed those plans, and as you know I joined the Navy Hospital Corps. My ever being a graduate student was dim that day.

But anyhow, there I was and I had one of those "this can't be real" sensations. Beautiful avenues of stately royal palms, stunning place. I smelled my way to the biology building; and looked into the labs. Sort of wistful, you know. My reveries were interrupted by a young woman, big glasses, big smile. And she asked if she could show me around. Of course, and she was great! She was a graduate student working on marine algae, a local girl of Chinese-Polynesian parents, married to a soldier, who was a graduate of biology at Hawaii. He had come from Chicago, gone to the University of Chicago High School. We had lunch. I wrote down her name and wrote her my thanks later.

Not too many years later, when I went out to Stanford I discovered that she was Izzy Abbott, wife of my later best friend Don at Stanford, with whom I made those trips on the *Te Vega,* to do marine biology. He's dead now, she's back in Honolulu on the faculty, an internationally acclaimed scholar. That visit certainly rekindled my drive to go back to graduate school. I got a real shot in the arm from that, an unreal interlude.

And of course no visit to Honolulu was worth it unless you got to Waikiki, and we did. There were two hotels on the beach, the Moana, which was for military, and the Royal Hawaiian, which was only for submariners. I was at the Moana of course. Big rooms, elegant space, no engines, no rocking, no voices, no stink, windows open, sea breeze, the rumble of the famous surf. Gorgeous food and drink. We went sight seeing. I took rides on a surf board, or tried to. Goes like hell! Spectacular beach. Walked all the way out to Diamond Head. Shops on Kapaliani Boulevard. Broad lawns and driveways, leading to the Moana Hotel. Huge banyan tree, I'll never forget it, growing rounder by dangling branch roots. That was the capstone to our brief brush with splendor. Nothing like Waikiki in the old days. It's not like that anymore, I can assure you. Sixty six hotels in addition to the original two.

We left Pearl by mid-October. And we chased the combat zone. We were always arriving in the forward islands after they were secured. Eniwetok, our first sight of decapitated palm trees. Guam, still mopping up. Ulithi, mines just swept. New Guinea, just secured. Manus, base being developed. Leyte, Philippines, still fighting in the hills. The western Pacific, that was our chase; we wandered for three months, escorting convoys. No vacation here though. We were sea-going veterans. This just happened slowly, day after day, steaming as before.

And now, the officers were known to me, and the crew. I had passed the tests of seamanship; in and out of strange ports, dry docks, fueling at sea, storms, drills, discipline. I was no longer "Mr. Bovbjerg, Sir," "Lieutenant, Sir," or "Captain, Sir." "Skipper" was the member-of-the-family label now, and that was wonderful for me, and it was reassuring to the crew even though I still looked like that college kid.

Steaming as Before: Life at Sea

SAMPLE LOG PAGE (PHOTOCOPY)

Log of the YMS 353 Attached to the

Naval District, Nov 15 , 1944

Hour	Wind		Barometer		Temperature		State of the Weather by Symbols	Clouds			Condition of the Sea
	Direction	Force	Height in inches	Thermometer, attached	Air, dry bulb	Air, wet bulb		Forms of, by symbols	Moving from—	Amount covered, tenths	
A. M. 4	NE	2994	1		79	55	Bc				
8	NE	3002	1		89	86	Bc				
12 M.	NE	2998	1		87	85	Bc				
P. M. 16	N	2998	1		86	84	Bc				
20	N	2996	1		84	82	Bc				
24	N	2998	1		82	78	Bc				

REMARKS

0000 - Steaming as before on Course 185° T - Speed 9 Kts

0640 - Course changed to 175° T - Speed 9 Kts

1215 - Changed course to 160° T - Speed 9 Kts

1910 - Course changed to 188° T - Speed 9 Kts

JMS.

Examined and found to be correct.

Richard V Boolgierg

5. Steaming as Before: Life at Sea

I sang a song as a kid:

> I joined the Navy to see the world.
> And what did I see? I saw the sea.

In the Pacific War we were not profound thinkers on naval purposes nor did we challenge the "Navy way." We went where ordered and we did what we were ordered. We were at war and our purpose was to defeat a distant enemy so we had to sail the seas. This chapter tries to paint the picture of life on a very small ship like the YMS 353, impressions and feelings. The key phrase of the chapter title is "steaming as before," day after day, the Navy way. We saw the sea.

"Steaming as before" is the inevitable entry in our ship's log while at sea; this log is a daily journal of the ship's major activities; no details. Thought you'd like to see a typical page (Take a look at the opposite page. This was the handwritten page—done on the bridge. Note the error in columns). We were en route to New Guinea from Ulithi Island.

The course changes are all that were deemed necessary in the log that day. Nothing happened! And this was repeated day after day. What this mundane account does not answer is the obvious question. What did thirty men and four officers actually do in that twenty-four hour period? We did our individual jobs and stood watches. The gunner's day was eating, sleeping, repairing guns, and two watches at the helm. The Navy says that there shall be four-hour watches, men and officers. There was bitching about duties and about watches, but no debate. This was the Navy way and it worked. Rules and routine eliminated the need for decisions. Day after day, "steaming as before."

It was boring. Tedious. Dull, dreary, monotonous. These are twenty four hour days. They merge, there's no start, no stop. Thirty five guys, same guys every day, day and night.

The ship is a mechanical speck on that big sea. The environment is tiny, it's crowded, it's hard, it's noisy. And your world rolls. It may roll very slowly, or with a chop. It may be a side to side roll or it may be a roller coaster, fore and aft. Or the worst, a corkscrew. And this is the whole world, compressed. The people are compressed; their souls are compressed. Their joys and frailties are magnified by this compression. It's no wonder that sea stories have been so powerful and so revealing of human nature. I certainly did think that as I looked around the little YMS 353.

And how we counted the days till the next port. As soon as we got underway we had an anchor pool, estimating the exact day and hour and minute till the hook was dropped. Anybody who wanted to could join it, and kick in a buck. And that meant that somebody got a $30 prize for being closest. I couldn't join, of course.

And yet, we could not wait to leave port. Going to sea was exciting after days moored in some dock or anchored in some harbor.

"Special Sea Detail, all hands!" Everybody rushed to a designated position. And the clank, clank, clank of the anchor chain was really music. It was great. And then, of course, the bosun out there on the anchor detail would holler up in the old mariner's language, "Up and down, sir" This meant that the chain was now standing straight up and down. The anchor was still on the bottom, but there was no slack in the chain. Theoretically, if the holding isn't too good you have to let out several more times the length of chain as there is depth. The anchor simply hooks the chain on the bottom. It's the weight and mass of all that chain which really does the holding, or so the book told me. Then "Anchors aweigh." That means the anchor has broken from the bottom and is now dangling straight up and down. And then, "Anchor in sight, sir," wherever the clarity of the water permits that. Finally: "Anchor at water's edge."

On the bridge, "All ahead, slow." And then the ship, which had been wallowing at anchor, or sort of straining at the mooring lines alongside the pier, is free. Then there's that little chuckle of water as we gain steerage-way and the water begins to slip by. Now a little wake at the stern, the bubbles and froth. The engine room rumbles a little bit louder. A smudge of smoke puffs out from the stack. Soon she has a bone in her teeth, a bow wave. Finally, "Secure from Special Sea Detail. Set the watch." There's something sort of joyous about that. "Set the watch!" The ship finds its way to the assigned position in the escort screen. At one point in upping the throttle a "critical speed" evokes a whining rattle to the ship, teeth chattering. Momentary. Then it's gone and the engines purr. And we're back at sea. How wonderful the sea!

The sea is the natural home of a sailor on a ship. The ship and the crew somehow shake free of the land. You tend to take deep breaths of air as the breeze goes whipping by. You leave the smoke and the noise and the stink of the land. All the compartments get aired; they all have funnels or scoops of some kind. The anchor is hosed down, and that usually smells, especially if its been in a harbor with a muddy bottom. The decks are hosed and cleaned with brushes and swabs. The canvas covers come off the guns; they're all aired out; they're oiled and they're checked; they are test fired. Somehow there's laughter everywhere, horseplay and shouting and glee. The scuttlebutt sweeps the ship. "Where are we going, what's coming up?" All kinds of scuttlebutt, somebody heard this; somebody heard that.

And somehow, we become closer. You're stuck on here, and it makes a family out of everybody. There are no strangers; there are no workers and no waterfront clutter, no packing crates all full of new gear that has arrived. The stink of mail is gone, with just the memories of those letters. It's so different and so wonderful to be at sea.

Then crash! Before the day is over, everybody wants to take a nap. You're tired, not used to the swing of the ship. Then an inevitable weariness overcomes you and this is what wards off madness. Everyone took naps and they would sleep anywhere! Put a life-preserver under the back of their neck and they're sound asleep, snoring. Sleep is indeed a way to thwart madness. You escape. Almost everyone is a bit queasy the first day out, but a nap is cure. And there's a raucous humor and teasing that also overcomes frustration and the weariness. But probably most of all it's the ship's duties that fill the daytime spaces—work. Naps fill in the cracks and everybody gets at least one good sleep. Even if they have the night watch, they get one sleep of several hours.

The moods of the ocean are so different. I only once saw the Pacific Ocean really pacific, glassy smooth, with only flying fish causing ripples. Even then there's a ground swell. This is something the landlubber doesn't realize. It isn't flat even though it is very calm. A ship miles away suddenly is clearly visible and then after a few minutes you can only see the top of the masts. At that point we say, "She's hull down." You can't see the hull, only the mast. The ocean has a bulge between you and that ship some miles away. And then after a while you see the whole ship again; it has risen atop that slowly moving bulge.

In storms we would disappear in a trough between the waves. The waves were higher than my eyes even up on the bridge, in typhoons. I estimated that we had sixty foot waves at that time. Usually we'd get sprayed even in just moderate chop; we'd get spray up over the focsle. In a storm, actual green water would run over us and crash; the bow would go down and this solid water would be sliced. The bow dug into the wave and flung the water up. It smashed on the bridge and then cascaded down the whole ship. Finally, the decks would have running water. There were scuppers (drains) all along the side and the water would rush out in gouts. But in any kind of weather at all, we were a wet ship. People didn't realize that the main deck, where all the work was done down below, was only six feet from the water. That is awfully close to the water when it gets stormy.

There's danger in a following sea, where the stern gets lifted, and if you're not exactly parallel with the wind you get that terrible corkscrew that can throw you off your feet. Sometimes we would roll as much as forty-five degrees. How much more we could roll and not founder, we never found out. But a large number of YMS's, I recall that a third of the YMS's lost in World War II, were lost in storms. We were so shallow; we drew six feet forward and seven feet aft. We were wood, no keel, and so we wallowed, we wallowed. A buddy from my previous ship lost his life in a typhoon on a YMS; it rolled over, turned upside down and sank.

There was a lot of challenge in those heavy waves. Many times I would go way back on the fantail. There were two depth charge racks back there and rollers that we ran our gear over. There I would stand and hold on tight. We always had a chain between the depth charge racks so that we wouldn't be washed over the stern. I could lean against that chain, hold on and then watch the bow dip down into the next wave and throw that water up. In a matter of seconds that water would come sluicing out of the sky threatening to drench me, and it did. This just made me bellow with joy! It was a pure, physical, youthful exuberance, conquering nature.

We would literally climb up the side of a wave in a storm, the bow much higher than the stern. We'd get to the top, and the wave would slip under us, we'd tip forward and slide down the far side. The bow was now much lower than the stern. As we went over the top the screws would come out of the water. They'd scream as they turned in the air. Then we would dip down and the screws would bite and we would shoot down the side, not only propelled by the screws but by gravity, like a sled going down a hill. We gathered speed until we got to the bottom and then we plunged. The bow would go straight into the water like a knife and it would shudder and everything stood still. We stopped moving; we were stock-still. We were in the bottom of the trough with a previous wave behind us and a wave ahead of us that we had to climb; we had to labor up that next wave. With that plunge into the next wave, the buoyancy of the bow would fling the water back. Then I'd get drenched another time. I spent hours back on the fantail.

Any kind of weather with tilting decks meant knocking off the ship's work. Guns can't be oiled under the canvas covers. Cooks cannot create. How about repairing the radar mast? Frying potatoes? Typing? We had one duty only: lash down everything that could shift. And of course there was always something that could still break free, "the loose cannon."

That kind of storm made a mess on the ship. Anything loose was tossed around like peanuts. The galley was a mess. The engine room was a mess. Everywhere there was a mess! Down in the engine room they couldn't leave a tool around. Even though they were working, they had to always hold onto everything. We got banged around in the ship. We got bruises and sore elbows, asleep or awake. This was wearing and depressing.

Did you ever contemplate walking on a rolling deck? It is an adventure. Cross the bridge? No, wait till the deck is level then run, really run, across the bridge and grab a pipe or something to hold on to. If you try to run down slope—crash! Up slope—no traction! Timing is everything. The cook waited for a level deck to go to the shelves across the galley. In the engine room locomotion was monkey-like, hand over hand on pipes and railings. The deck was a metal grating, which helped. A day of this and you were exhausted; it took enormous agility and left a tension.

Eating was a chore. In our wardroom, we had a small table. Four people could sit around it, like a booth almost. We had about inch-high railings that had pegs in the bottoms to fit into holes, so we could criss-cross the table into compartments. And in front of each person there was a compartment that would just hold a plate and another one that would just hold a sauce dish or a cup. Then things would only rattle around inside these; we didn't have to hold onto everything. There finally came a time when we could not eat meals, and just had sandwiches.

The officer's head was tiny; small shower stall, sink and toilet. We had fresh water, but cherished it. Standard shower: get wet; stop water; soap down; rinse quickly; stop. But rolling ten or twenty degrees first one way, then the other? Corkscrewing? The water pipe was the only hand-hold but the stall was tiny, elbows could be wedged. Shaving? One leg wrapped around the sink leg, one hand clutching the window port latch. Sitting on the toilet? The water level stayed still while the ship rolled. So water went up the back then up the front of the toilet bowl. Did it ever run over? Lots of times. Now contemplate the unholy douches at times. A mess? Yes! An adventure? No!

Now imagine the simple task of taking a pee. You could keep your balance with great difficulty, but then the toilet tipped. If you wedged between the door and the shower stall you stayed even with the toilet but the stream bent! The stream after all could only acknowledge simple gravity.

And then all the time at sea there is this swirling stink of diesel exhaust. Even today if I get behind a bus going up hill, I see this blue cloud and am assaulted by that diesel smoke and I'm swept right back to the sensation of being on a rolling ship. When we had a following wind the same speed as our forward speed, we lived in an unending cloud. Headaches, an unbearable situation.

At sea is no place to be if you want to get something done on a small ship. There's no energy left over, just weariness. You're broken down somehow. You try to sleep. We had sides on our bunks, very lovely, wooden bunks in our little cabins, our state rooms. I slept, always, spread-eagled on my belly or on my back with my knees, ankles and elbows touching the side so that I could always brace one elbow and then the other elbow, this knee and that knee. I did this while I was asleep and didn't fall out. But it left me tired when I got up.

There's no way to study. I tried. You can escape in some paperback books or magazines, or you can play solitaire in the wardroom or cribbage, have bull sessions, private times in the bunks, re-reading the letters, shower, shave, sew on a button. Anything done was accomplished the hard way.

This was not just the Navy or the war; I've been on four oceanographic trips since the war, on well-appointed but small scientific ships, and I had the same feelings. It came right back to me. We didn't get anything but routine done when we were at sea. When the ship had tied up or dropped the anchor, and the ship was stable and quiet, you could have some energy left over; you could do creative scientific work, but not underway.

In the South Pacific, near the equator there's not really any difference in temperature, winter and summer. Hot all day, hot all night. The only real wind frequently was the 9 knot speed. The humidity was one of the terrible things. I recall opening a locker in my stateroom where I kept my dress blues. When we came out it was fall; I had dress blues on. The humidity was so great that when I opened that locker all my clothes were bright green. Bright metallic green. This was mildew. It was just terrifying! Shoes were bright green. I frantically took the uniform out onto the deck and beat it with a broom, scrubbed it, brushed it, and in just a few minutes it was just as blue and nice as it had ever been. But I knew that fungus was getting a living off that wool, which meant that it was degrading the cloth, slowly disintegrating it. But very slowly. That's the measure of the humidity. Everything was rotten and smelled of mildew. There was no escaping it.

Even in rough seas our weather in the Pacific was often stunning, tropical seas and skies. The sea sparkled and the backdrop to this stage was the vast dome of sky. Sometimes there were no clouds—rare; sometimes high flecks of clouds, a mackerel sky; sometimes towering thunderheads, awesome; sometimes story-book fleecy clouds all around giving welcome bits of shade for a minute. But weather and the sea were our life. We went on deck (at least I did) all the time just to check. Sailors are probably more aware of the sky than farmers.

On most days the ship's work could go on; every man aboard had duties. It was routine, never questioned. The duties were almost always self imposed. The yeoman knew it was time to get the pay roll files in order. The electricians had lists of jobs to be done. I surely did not tell them what to do. This is a small ship attitude, no big table of organization. Of course it was boring. But there was pride in it. And one could do no less; the Navy way, routine.

And of course we had to "fight the war." The convoy we were escorting had to get through unharmed, no sub attacks, no plane attacks. We went on watch, off watch, on watch. No subs, no planes.

A biologist also got his kicks at sea. I saw things I would never have seen. I also saw a thing which was very surprising; there isn't much life in the middle of the ocean.

We saw so many porpoise. They would suddenly pop up at the bow, or come alongside. They were always "smiling;" their face is structured in that perpetual smile. The amazing thing is that these are mammals, of course, not fish. They have mammalian eyes and not fish eyes. In those eyes you see this, if I may say, awareness, the way you see it in a horse or in a dog. It's a knowing eye. And I have looked down, and the porpoise looked up at me, and there was a contact the way it would with a horse or a dog. It's a surprising sensation. And there they are, they are so damn graceful and so fast. They seem to aggregate at the bow of a ship. They would plunge along at the same speed that we were going, leaping up and then making those gorgeous loops into the air.

One day off the coast of Costa Rica, coming up to San Diego from Panama, I heard this enormous noise. I ran out on deck and I saw a giant manta ray. This giant thing, that would fill a living room, undulates its fins (it's flat and diamond shaped), gathers speed, and then forms a plane going up. With this speed and that flat plane it would clear up into the air gliding, and then land with this enormous wallop! To simply get rid of some of their skin parasites this way? But it was an incredible thing to see. I'd seen skates in shallow water, the stingrays, but nothing like the giant manta.

We saw whales quite a few places in the ocean. We'd see the spout, there'd always be a holler. A whale at sea is always an event. Sometimes we saw them close; usually we saw only their spouts.

Jellyfish. I saw some as large as the kitchen table. And then there were clouds of little ones the size of a thumb-nail, tiny little ones, but dense, especially along shores.

Flying fish, wonderful in the tropics. I spent a lot of time examining those and I had read about how they fly. They don't fly of course, they glide; they leap clear into the air and then scull with their tail. Sometimes they landed on our deck; we were that low and they could go that high off a big wave. Some ended in a frying pan.

I saw a surprising thing in the Philippines, in the Mindanao Sea. We were right behind a typhoon, just missed it. We could not go through the Mindanao Sea without continually bumping into something. We hit branches, whole palm trees, and some piles of trees, all interlocked and tangled by the storm, then driven out into the sea. It became clear to me that what I had learned as a student in pure theory, the rafting of species from one island to another, was real. A lizard, a snail, a beetle, a fern, a mushroom, ants, would be living on those logs. If they were driven to another island in a matter of days, here would be a population starting anew, having been driven by typhoon waves and winds. That was a remarkable thing to see and I would never again doubt the efficacy of rafting as a means of gene flow, in biologists' terms.

I brought some books with me: Sverdrup, Johnson and Flemming, *The Oceans,* just out, and a bible of oceanography; and Hesse, *Ecological Zoogeography,* I'd had a course reading that; and then Buchsbaum's invertebrate zoology book. I felt that I was more a biologist. I kept telling the crew how lucky they were to see all this stuff. Unconvincingly.

The men occasionally went fishing. They trolled for bonita; caught some that ended on our plates. We caught sharks; some excitement. But the upshot of it was, as Hesse said, "Blau ist die wusten farbe des Mer" (Blue is the desert color of the sea). The open sea is empty, biologically speaking.

With all of my emphasis on how boring and endless the days at sea were, how weary we became, how stale, and how little we felt we could get done of any substantive nature, the real job did get done. We did this job of sailing a ship. That is what we were there for. The long days were punctuated with something that sparked the day. Maybe a radio story on our victories at sea, or the progress of the war in the Europe. Usually the news was dull. There were all sorts of little events and emergencies. Small ones and not so small ones. Engines always had breakdowns. And then there were crew emergencies. There was bickering and there were problems. And of course the enemy relieved the boredom, until that too was wearying. Mostly the enemy was false alarm. We'd get a bogey contact. The commodore would radio: "All hands should be on the alert. Bogeys seen to the south." Then came an all clear. We'd secure from general quarters and go back to our naps or our jobs.

Rain squalls came and went. We danced with water spouts. And in so much of the area, storms were visible all about; little clouds with rain creating a silvery veil below. And this is so true of the tropics. All of a sudden we'd be in it, and then the deck would be converted to a bath, with soap, bare-assed sailors joking and scrubbing.

There was ship handling and convoy maneuvering to check somnambulism. Suddenly we'd get a coded message; the skipper and the officers would decode this in the wardroom. There were constant convoy communications. We all had signal lamps; lights would come at us. "Dah-dih-dah-dih-dah." And then our call number; we would answer, and then would come messages from the commodore, something we should do. There were flag hoists for course changes. The flagship would raise the flags at the dip (half way), and then every ship in the fleet would raise those same course change flags to the dip. The flagship would raise the flags to the peak; everybody would raise flags to the peak. The flagship lowered flags and all ships would follow. This was kind of old fashioned communication, but for course changes flags were great.

Then we had TBS. The short wave radio, called "Talk Between Ships." And this could be heard only to the horizon, unless there were strange skip waves, which there sometimes were. But this was on the bridge, and was always monitored, someone was always listening for the YMS 353. I remember for a while we were "Merrimac 353", for a while we were "Lena Horne 353." No one could ever figure it out.

Nights were very different. No duties, just sleep and standing watch. The official day at sea began at midnight, when the page in the log was turned. Before midnight the galley would come alive with the new watch going on; they'd all grab a cup of coffee and they could make a sandwich. But the rule was that fifteen minutes before the watch you were ready to go, go out on the deck and look around for a bit, get the night vision. And in the case of the officers they'd go up to the bridge, and say, "Hi, what's up."

And the officer of the deck, tired, couldn't wait to get down to his bunk, would respond, "Well, we had one course change an hour ago." And then they'd chatter a little bit.

"Any new orders from the captain?"

"No new orders, just the usual captain's night orders"

"Okay, I relieve you sir".

"Aye, aye. I'll see you later."

Below in the engine room the same scene and words.

And that would be a new watch. That was so long a time from midnight till 0400 That was not a very good watch. It's so lonely. You'd get a nap after supper, and then be awakened before you really have had any sleep. You'd be relieved at four o'clock in the morning, and know that you only had a short time till chow. Then work to do. This was a bad watch; it was a very bad watch.

It was quiet on deck at night. On the bridge it was so quiet. In the engine room, the clatter and rumble of machinery. There was not a lot of chatter. You just did your job surrounded by the sounds of the waves, wind, engines, and total darkness. On the bridge the silence was punctuated by the sonar ping, sonar ping, sonar ping. Radar didn't have any noise.

This darkness was a blanket, no lights anywhere. All the ports, and hatches were dogged down, just absolutely light tight. When the hatch would open to the wardroom, all the lights would go out and a red light above winked on. Then they'd slam the hatch shut, turn the dogs; lights would blaze on again. It was total, total darkness all over outside. A match flame is visible for three miles we were told.

You couldn't see the sea. You could feel that it was there, you could hear it hissing and bubbling. Land was probably thousands of miles away. Other ships, even with good night vision, were still only dark lumps. On a dark sea there's no horizon. Where the stars ended that was the horizon. On the other hand, dark was cozy; we felt very safe from enemy planes.

During bright moonlight, we felt naked. Ships were outlined on the horizon and had a sparkling backdrop. We were great targets. On the other hand, submarines never were good targets, and planes, even on a moonlit night, could come in behind clouds and make great approaches. It was not in our favor.

To me the stars were a comfort. I'd had a hobby in high school that made me very familiar with the night skies. A friend (the guy I visited in

L.A) and I made a telescope, which we had up at our summer camp, where we were counselors. I used to go out at night with the boys, and by myself, set the alarm clock, sleep out on the hill where we mounted a six inch reflector telescope. We spent weeks and months grinding the glass and getting a metal tube and all the necessary bearings for turning it on a gimbal that would match the latitude so we could point it at the North Star and turn from there. It was really wonderful. It got us interested in astronomy; we were familiar with the "movements of the spheres." So at sea when I would scan the sky at night all my stars were there. I could tell that Vega was almost down, same as last night. I was shocking the crew with this erudition. They were sure that the skipper knew every star by name!

I was curious to learn the new constellations as we went south, and when we crossed the equator, it was fascinating to leave the North Star behind. At the equator you look north and the North Star is right on the horizon. You go farther south and it's below the horizon. The Big Dipper, which rotates around quite close to the North Star, would arise, make a little short circle, and set. As we went farther and farther south we lost the northern constellations, but got new friends.

The bioluminescence was shocking. Our wake was a solid, pale silver green. It just pulsed with blobs of emanating light. I was fascinated. Sometimes I could see huge jelly fish—sort of boil and turn over and their body was blue lined. I couldn't catch anything in that glow because it was all too messy, gelatinous plankton. That would have to wait twenty years, when I went back to sea as a biologist studying plankton.

On the bridge the radar scope was a pale luminescent screen, like a TV screen. We were a blob in the center, and a bright line, a radius, made a slow circle, over seconds. That was the sweep of the radar dish up on the mast. And as the other ships in the convoy bounced back this radar beam, they were suddenly revealed as a bright light on our screen. A large ship was a larger blob of light than a small ship. Then they would fade as the sweeping line went around; then the next sweep would pick them up again. And so we could say, "Well, now, there's the flagship, it's 2.1 miles away (nautical miles). And it is bearing (there were marked relative degrees,) exactly to starboard." We had a bearing and a distance and therefore we could keep station very, very well. The ships at the top of the circle were ahead of us, the ships at the bottom of the circle were behind us. And we had scale rings that we could shift and the settings could be changed. I remember seeing

New Guinea at 60 miles. Of course, that's because there were such high mountains, the Owen Stanley Range, as I recall.

Radar was very useful in station keeping. Frequently we did maneuvers including zigzags, where we'd have frequent course changes. And when that happened and the flagship suddenly turned toward us twenty degrees, we would have to turn farther out and slow our speed to drop back into our assigned place. And so there was always this sort of ship handling that went on in the middle of the night; with all lights out, radar was just a godsend. One glance would show the flagship and if we needed to be 400 yards farther to the right we could just move over until we were right on, till it was exactly the right bearing and exactly the right distance. Then we knew the course and speed and we would just stay there; occasional corrections.

On a few occasions the radar would go out and we did not have a spare tube, or something was broken. It happened once in the Philippines, in Surigao Straits, a very narrow place. We were escorting a convoy in total blackness and we couldn't see a thing. We couldn't see mountains that were just two or three miles away. We couldn't see the other ships. It was terrifying! It was like the good old days. We'd doubled the lookouts, and we could talk to the nearest escort by TBS, "How close are we, how close are we?" And they'd call us, "You're getting too close." But it was terrifying, that trip through the Surigao Straits below Leyte Gulf. It's not one I'd like to remember.

On one occasion I thought the course was going to take us to the top of the mountain. I had to call the commodore, and got chewed out for not knowing where the mountain was. You don't excuse yourself that you don't have radar. You shut up.

The sonar was different. This was not the silent ghostly glow. This was a shrill sound! It was also a screen but it was the sound that was pitiless; this constant racket. It's the same principle as the radar, but it's underwater sound which bounces off the target and returns; it's recorded and time is the distance; very obvious. And the direction is noted by the little transmitter down below the ship, which we could lower and raise and turn. This sound would go, "Sproing-ing-ing-ing-ing." Fading away. We'd usually ping in one sector, from dead ahead to the starboard beam, for instance. All the escorts together covered 360 degrees.

But we never had an enemy target. It was futile; we were in an empty sea. This was not the case when I had been on the Atlantic off Norfolk. We had German submarines around and we'd get contacts, the enemy, but on only a few occasions we'd hear this "Sproing-ing-ing-ing-ing, PIP." Suddenly the officer of the deck would rush over. The operator would twist the dial, sharpen it and hold right on it. Then if it were an emergency, call the Skipper. Probably a ship we did not see, one of our own. Or there would be underwater sea mounts not on the chart, a whale, even a school of fish. Good sonar men knew the sharp pip sound of metal as distinguished from, say, a coral reef or a whale. Then of course we got up and down dopplers. If the sound would rise that meant we were closing on the target, getting closer. We never had a sub contact in the Pacific. One time we went out and played hide-and-go-seek with a friendly sub for practice.

The 4 o'clock to 8 o'clock watch in the morning was really deadly. No one could get up at 4 o'clock and be in top shape, even if they'd had several hours of sleep. It's just the worst hour for anybody to have to come on watch. And there was nobody in the galley; there was a left-over mess from the previous watches. It smelled of peanut butter and stale coffee. Maybe there was still some coffee which was almost sludge. It's just an empty ship on an empty sea and nobody wanted to talk.

The bright part of that watch was dawn. There was this wonderful birth of a new day, a new virgin day. The east would pale just a little bit and then pink and yellow. One by one the rest of the ships' outlines would show up on the horizon. We could see the rest of the convoy or fleet. Morning after morning I had that same thrill of dawn.

"Pass the word, general quarters, ten minutes." We always had general quarters in the forward areas. Dawn and dusk were the favorite times for enemy attacks. And then, "Call the Captain for dawn sights." I almost always went up for sights on the bridge, at dawn. When the major stars, the first magnitude stars, were still out, there was now enough light to get a bearing to the horizon. Then we would smell the coffee and pancakes. It's just a wonderful time, just wonderful! And our own biological rhythms sped up and the ship sort of quickens. We could hear curses and laughter, and the day's duties would start, and the routine.

The sun rising and the sun setting were not really the same sun, or so it seemed. Dawn was so exciting; the long dull night was behind. Then there was the first sliver over the horizon. I had to keep reminding myself that the sun did not rise; the horizon tipped below the sun. It was so pale, almost pink against the lavender horizon. It seemed so fresh and crisp; new. That also is how I felt at that moment.

The setting sun was tired, like the sailors. It usually was a ball of orange, flame-like. Then, as the horizon claimed the sun, the water reflected it in an orange streak. The sun had not set but had melted out on the sea in a molten stream. In the tropics the sun sets straight down, no angle. So, sudden nightfall.

Sunset was invariably followed closely by two requests of the officer of the deck on the bridge. The cook would amble up, his only contact with that distant seat of power, and ask, "Sir, permission to dump garbage?"

"Very well."

About the same time a whistle came from the voice tube on the bridge. "Engine room: permission to pump bilges sir?"

"Very well."

So as night fell, buckets of garbage were heaved over and the bilges pumped free of leaking sea water and spilt fuel and oils. Were we ashamed that someone would see the string of boxes and potato peels? Or the sinuous line of oils and dirt? We never gave pollution a thought! We were afraid the enemy would see the evidence of our presence and maybe find our course. Time honored practice, every night after sunset.

All refuse went over the side at sea or in port. We did not "dispose" of a broken wrench, we tossed: "The deep six." Of course the wrench sank and ceased to exist. This cowboy mentality was accepted. The major ports became very polluted. Please remember how different times can be, for so many attitudes.

Captain didn't stand watches but let me tell you, he slept lightly. Any change while I was asleep instantly had me wide awake. For instance, if we changed course, that would bring on a new kind of roll. The ship would respond differently, it would be a different angle to the waves. Right away I'd be awake. Or if the engine sounds changed. If one engine was shut down, "Help, we've got a bad engine". I was on my feet in a second. Or if suddenly the engine was faster. "What's going on out there that we're speeding up?" (The Navy considered the Captain to be on duty and responsible day and night—on the bridge or in bed.)

There was a steering motor, an electric motor, that turned the rudder. Wires went to the bridge and the wheel, it was a great big wheel, but it spun beautifully. As it turned it made contact, and every half degree of turn there was a click, a solenoid in the electric motor. One click moved the rudder half a degree. On a quiet evening watch there would be a single click, then a long time with no course change, the wheel steady. If the helmsman saw a slight yawing off on the compass, we would hear, "click, click, click." Slowly the wheel turned, that was normal. Didn't waken me. But when I heard, "clickity-click, clickity-click" that meant the man had spun the wheel up there. "Man, why would anybody be spinning the wheel up on the bridge?" Oh, I was out like a flash!

I slept in my white skivvy shorts like everyone else, but I had right next to my bunk, a pair of Mexican leather sandals, *huaraches,* woven leather. I could jump into those and my feet would slip right in; the woven leather grabbed my foot. I went, pad, pad, pad, pad, up to the bridge. If there was nothing going on I'd say, "How's everything?," casual-like.

"Well, they called a quick course change on us and we didn't catch it fast enough. You probably heard the wheel."

"Yeah, I did."

And there'd be times when I'd dream this. I heard the wheel whizzing around. "What's happening, what's happening?" I'd jump into my *huaraches* and roar up to the bridge. There was nothing in sight anywhere.

"How's it going?"

"Oh fine, Skipper. Can't sleep?"

"No, can't sleep. Just wandering around. . ."

They thought I was awfully solicitous but a lot of those times when I showed up in the middle of the night on the bridge, it was because I was having a dream, that something was going on on the bridge that I didn't know about and ought to know about.

In that hour after sunset and before dawn and at noon were the times when we got navigational sights. I routinely did it. Since the other officers stood watches, the least I could do, should do, was get the sights. On the other hand, they all took their turns at it so they would keep their skills.

I really got good at it. I never gloried in the trigonometry; you don't need to know the trigonometry. I had a gut feeling for the concept of celestial navigation. I knew the stars and I knew their path circles. In fact, when clouds covered almost the whole sky but just a little patch, and I saw a bright star there, I knew that at that time of night, at that place, that was Sirius or Aldebran—I knew the stars. It's where it should be at 5 o'clock in the morning. It looked like that star; it had that yellow look, that blue look. You just had this gut feeling when you knew the stars the way I did.

So I made all kinds of navigational fixes when no one else could possibly do it on almost completely cloudy nights with only a few clear patches. Ours was often the only fix turned in to the flagship.

Let me digress a bit to talk about navigation. What I say will be grossly oversimplified! And this is navigation of fifty years ago. A hundred and fifty years ago? Celestial navigation was only part of it.

When near the coast we did not need stars; we used landmarks on the chart and on the horizon. We took bearings, the angle to a mountain for instance. We had an azimuth circle mounted on a repeater compass; these had a sight for aiming. If we could get three mountains we would have three converging lines to put on the chart. That was a fix, our position. Simple.

At sea, far from land, we kept track of our position on a blank plotting sheet marked in latitude and longitude. This was a working sheet. Each day we then recorded our position on a regional chart. We drew a continuous course on the sheet. If we were running 090 degrees at ten knots then we could draw a line due east ten miles long. This was our deduced reckoning, our ded reckoning (this usage from a scholarly navigation instructor—most folks spell ded as d-e-a-d reckoning!). This was a rough, tentative line. Was the estimate of speed accurate? The steering? No currents? No wind?

If we had a few stormy days, this line could be off badly. I once missed an island by twenty miles after three such days. Scary. So we had to correct and reset the D.R. line with celestial navigation. This is a formidable mathematical procedure that produces a line of bearing on a celestial body rather than on a mountain. And like mountains, it is better to have lines that cross from two or three stars. We were at that point where the lines crossed or made a triangle. A fix. That reset our D.R. line on the plotting sheet.

The classical old instrument for shooting stars is the sextant. It measured the number of degrees from a star to the horizon. It was hand held, had a telescope for sighting the star, and an arm that moved on a circle; it was calibrated in degrees and minutes and seconds. The arm was moved until mirrors superimposed the star and horizon. The degrees were read off and recorded. The exact time was critical. In the chart house we had a chronometer, a super clock set at Greenwich time, zero longitude. We checked the chronometer each noon to the second, via Navy radio.

Back in the chart house we took the sextant reading and exact time into books with tables on any star we shot. We had mimeographed work sheets for filling in the blanks and doing the arithmetic. We came up with a line for our plotting sheet. It was actually not mathematical—rather just some arithmetic; fill in the blanks.

These days it is so ridiculously easy. A gadget on the bridge has a digital readout of latitude and longitude, a continuous fix, to the second. It does this by electronic reading of the helm and speed. Then, every twenty minutes a celestial fix is sent into the gadget by satellite. I have been on oceanographic vessels with such. Oh my!

The actual shooting of the stars with a sextant is a bitch. With any sea running, one must wrap a leg around something and then make the body into a shock absorber like a slalom skier. Bent, soft knees. Some folks can get this knack, some can not. I did, but then I could read a newspaper on a bucking street car without holding a strap!

At sea—any time—some men caused trouble. They got put on report. Someone goofed, and whoever was senior to that person would say, "Captain, I'm putting this guy on report. He showed up a half an hour late for his watch last night. And that's happened two or three times lately. Something has to be done about it." That was official. When someone was put on report, it was official. The captain had to take notice of the fact that somebody was goofing off. And then he had to decide how serious it was.

The captain had complete authority. He could call a general court martial and this would require high ranking officers from other ships or a base; we couldn't do it at sea. They could give the death penalty. There could be a summary court martial, which was held aboard the ship, with senior officers, and it would give lesser sentences. It could be a deck court, just the ship's officers, for minor offense. Or the events could be so minor that it was just handled by the Captain himself. A Captain's mast was called, where the Captain was prosecutor, defender, jury, judge and appeals. Absolute authority. I preferred the mast to anything else if I possibly could get around having a more formal court.

Punishment was immediate, that was important. I was also able to make the punishment fit the crime and the man. I did not want to dock pay for a married man (I don't know if anybody figured that out). Most importantly, there was always instant judgment rendered and if there was culpability, some kind of punishment. The word got out around the ship that this was so and sometimes they would say, "Oh, the Skipper was too easy on the jerk." Or, "Did you hear what the Skipper made him do?" But the mast was on record, the yeomen took notes and made entries in the man's record; it was logged in. I was absolute master and I am sometimes frightened to think of the power that was vested in me. This changed after the war.

Sometimes I did actually hold the mast out on deck, "before the mast." I read or made crew members read the appropriate sections of Navy Regulations. They would stumble through this formal language. Some of these guys were really very poor readers. "Shall not strike or disobey a senior in grade or rank. Punishable by death (The United States then being in a State of War)." I had one young man read that to me a hundred times. No, No, No, he read it to me about ten times and he read it ninety times to the yeoman who recorded the number. He was quite chastened by that, choked on reading the word "death." It didn't cost him anything, but he had a record of being at Captain's mast and it said, "Appropriate punishment was assigned by the Captain."

There was a man who was a Bible reader when things got tough. He was reminded that "pusillanimous behavior" would merit the same punishment (death). He understood that. There was a way we could handle this. He was a very religious man. He carried this little Bible all the time in his hip pocket, in his dungarees. And when the enemy was almost on us, and I could see them, he would abandon his radar position and sit on the deck of the bridge and start reading that Bible. He crouched, moving his lips, sincere, huddled in his life jacket, helmet pulled down. We had an agreement. He would not quit until it was perfectly clear that he'd given me all of the bearings of the enemy on the radar until they were close; until I could go out on the side of the bridge and see every one of the things he could see on radar. And he said, "Have you got all that, Skipper?" And I would say, "OK." Then he'd abandon his post, sit down on the deck and start reading the Bible. It never got in our way. I never punished him, he knew the regs. Not all the officers agreed with such light admonition.

Sometimes the Captain was told that somebody was put on report and I would talk the chap out of it. "Oh, you don't have to put this guy on report, let me talk to him." Man, the talks I had in my cabin with tense, young, frightened, bored, frustrated, away-from-home guys. I was not one of the older guys on the ship. But I was the "old" man. And that gave me the authority to talk to them in an old man way. They would talk to me about the terrible missing of Mom and this kind of stuff and wondering what was happening to old Fanny, his, his hunting dog. These led to problems in the crew, so that somebody would put them on report for doing something wrong. But the Captain's mast worked; so did the talks.

There was one very special case that I have looked back on, amazed. The most gross injustice was the utter segregation by race in the Navy and indeed in the nation. We had one so-called, "Negro steward." He was the officer's steward, the gentleman's gentleman. He would open a slot into the galley and the cook would shove food through and he would serve it. Go around and ladle the soup in our bowls and pour coffee.

He shook out the tablecloth before meals. Once a week he washed it and all the linen napkins, ironed them, ironed them in the galley with a hand iron. He washed our clothes. There was a steward's day on the laundry machine. And he ironed our shirts. He repaired buttons. We were his entire duty. He cleaned the wardroom, swabbed our deck. He cleaned the head. And waxed the deck so that it shone. He polished everything. He was a servant.

This was pure slavery. He was a New York cab driver. He was street hard, not very big but tough as nails. But so sweet. I never discerned any question in his mind about his job. And sometimes I got the feeling he felt lucky he didn't have to do all this crap that they had to do in the engine room or out on the deck. He had a slow grin, teased with the other guys. They liked him and respected him for being on a gun crew, but he was really accepted for his own character as well. I felt close to this guy. He talked to me about his home life, cleaning the deck while I was doing some work in my cabin. How close were we? We weren't really close. I know that now. It seemed like we were close. Was I just a kindly slave owner? "We treated them well," they used to say. "How could they possibly resent being slaves? We loved them." I don't know, but it's scary to think back on it. I remind you that segregation was tight back home: strong anti-semitism all over, and in the South, drinking fountains with signs above them "colored" or "white".

The rest of this chapter, "Steaming as before: life at sea," really should have a separate sub-title: A Helluva Mess on the YMS

You sons who sat at the table in a family with biologist parents will see the point to this riddle. What is the same about these six places: Prisons, zoo cages, city slum tenements, a round bowl with four crayfish, an island of nesting birds, and a YMS? The too obvious answer is: crowding.

But what comes from crowding? Spacing, territoriality, peck order; internal and psychological stress, violence, social breakdown. A biologist can quickly point out that the YMS has other problems. This is a population of young males with raging hormones, and no females. This is a legal caste system with no grievance system. And here lurks the face of death, by the enemy or the elements.

That is enough!

So how did we ever make it?

We could find outlets for the frustrations and venom: we could have absolutely routine activity, work. We could have acceptable rules. We could turn to each other.

These things we did. And, as I write this, I can hear you saying that that sure sounds like good old Dad, the retired professor! You are correct. No one worked out this cause and effect then. We had the Navy way and we had intuition. We just lived on the little tub day after day. That is what I am trying to say in this chapter: "Steaming as before: Life at sea."

But even at the time, I mulled on this idea that has been suggested earlier: the small ship is an intense microcosm, a compression of the world onto this small stage, all the good and evil, hope and fear, hate and love. A compression of all of these into latent volatility.

But enough of this academic stuff; what happened when the kid skipper was twenty-five years old? My intuitive response to the tensions were of two kinds.

My first response was sticking to Naval traditions. I really believed in that and it was my duty as captain and I would not dream of avoiding it. Always be formal on the bridge and doing ship's business: nothing sloppy, no kidding; give orders in the traditional form. The Navy way works only if it is routine and expected, unthinking; then no problems and no difficulties. Lots of officers, often reservist, leaned on Navy Regs too much. This was often youth and lack of confidence. Then there were others who just could not take the job seriously. They saw the Navy way as unfriendly, out of date, and undemocratic. They wanted to be buddies with the men and have their confidence. This would be okay except at two times: when meeting the enemy and the unkind elements. And these two are what mattered most to the Navy.

But I also intuitively felt that there had to be a second attitude that would be a safety valve to the formality. And a safety valve would surely be needed on that helluva mess on that YMS. This attitude was informality and joking when not in the line of duty.

I really did get along with the men; we had conversations on deck. This usually led to kidding and laughter. Sometimes we talked shop in an informal way. And I did not require uniforms. The officers did not need badges of authority. Not hard to keep track of four officers! There was no saluting or standing at attention. Enough to worry about without that. And it was in the tropics; the less we wore the better. Officers had khaki shorts, white skivvy shirts; I wore Mexican *huaraches* on my feet and a pith helmet on my head. The men; dungarees and skivvy shirts and maybe an olive drab baseball cap. Goodness sakes, what laxity. Great! This was part of the informal attitude of the skipper, who had the authority given him by the Navy.

The word got around they could always come in for that chat with the skipper. I heard some serious things; I heard some ridiculous things. "My wife wants a divorce; she ran into this guy at work." Or: "My brother has been sent to the Marine Corps prison." Here were two serious ones I remember. One guy had a serious neurosis. For another one I had to go to a chaplain; I made arrangements for him to talk to my guy and get in touch with Red Cross to see what was going on back home.

But I could joke about the silly stuff and they would come in and talk to me and we would laugh about it and they got a release. We had a tough, little ship, but basically one with happiness. Without going back to the old Navy saw, "A taut ship is a happy ship," we were a taut ship in the formal sense of doing our duty, using the Naval traditions. But we were not a taut ship in the way we lived our lives and the way we talked to each other, off duty.

Looking back, may I pull together what it was that made me what I was at that time. I was such a combination of things. I applied Regs, Navy Regs and traditions, but I also was a low-key kind of a guy; actually, I was a terribly gentle person and I think they knew this. Don't forget I had a family with traditions and a society with traditions and this emerged. And we had lived through a depression in those days. We knew hard work and poverty. We accepted authority; we had worked for a boss. I'm talking civilian life. Getting the job done; this was in my tradition, as it was for most of the crew.

All of us were used to competition. We'd been in sports; winning was everything. Truly, I mean life was winning. This was a very prevalent male tradition. There were unwritten manhood rites. Honesty. Taking one's medicine. No snitching on your fellow crew members. Let off the steam with joking and tussling, cursing, laughing. That these were all means of releasing stresses, goes without saying. You must retain your sanity on such a small ship. The Navy can only give you the framework for this.

Perhaps the fault that was most egregious in the view of anyone on the ship, was not pulling one's weight, or being chicken. It certainly held for me. I intuitively knew that I was the one person everybody looked to and at. On the bridge my voice never quavered. I guarded against that. I had awful indecisions inside, but would never let on. Most of those guys thought I was absolutely made of ice. Far from it! In battle I intuitively went to the side that was exposed; this was in fact the only side to see what was going on. If there was a possibility of being shot at from the starboard I went to the starboard side of the bridge, to see. When a plane was diving on us I went to the side where the plane was, to see. I don't think I thought about that. If I had, I think I would have come up with the idea that it doesn't matter where you are on the bridge on a little ship like that. If a *kamikaze* hits you, it's history. But I also felt the need to show this, that I would put myself in harm's way.

I never hollered at guys in anger, only to make myself heard. Always used the low voice, and the correct phrase on the bridge. This was not bravery; I just could do no less. I was programmed. All these forces were working on me, as I look back. But this was the kind of guy who was never late or absent in grades kindergarten through 8th grade, not once, not even for a cold! Or a blizzard! They'd awaken me in the middle of the night. "Skipper, the Officer of the Deck would like to see you on the bridge." I was instantly awake; never snarled, never snarled at one single guy. I'd say, "Thanks Don. Tell the OD I'll be right up." I was programmed to be nice to the messenger. This was unvarying; it was built into me. I truly believed that manners were the lubricants for social machinery. I know the crew talked about it. Sometimes it came back. "How can you always be so nice in the middle of the night?" They got the other all the time!

In many ways I was lucky to have the outlook of a student, interested in lots of things. This was part of my university experience but also my home. Your granddad was a student when he died. The house was full of books which had been read. So what did I do in the Navy? I read the Navy manuals. I needed to know this stuff. I read the routine reports the Office of Naval Intelligence sent out to commanding officers, confidential information on global war, politics. I spent time with different gangs, the black gang, the cooks, and we talked. I read the ship's book from BuShips and the Bureau of Ordinance. In the *Mine Identification Manual,* I inserted all these pornographic photographs I got on Kapaliani Boulevard in Waikiki Beach. You know, like the two-horned type, the magnetic type, the acoustic type. These were big glossy photographs in their full frontal nudity. No one ever saw those outside the wardroom.

Storms were fascinating to me. How high are the waves? Why are the sunsets different at one part of the Pacific from the other? I really sat around thinking about these.

One of the things that was so fascinating, I must confess, was profanity. Profanity takes over in the absence of any ameliorating social forces. Apparently, it's a release of some kind. Here is where the word "fuck" is used as every part of speech. It's a noun. And somebody is a "mother fucker, a dog fucker." It is a verb, "somebody fucked something up." I remember one time a bosun's mate was hammering the threads of a bolt with a cold chisel so that all the threads were jammed down on the thing. It could not possibly be gotten off now without a blow torch.

"What are you doing?"

"I'm fucking them threads."

And it was the fucking engine, it was the fucking radar, it was the fucking coffee. Or an expostulation, "Oh, fuck!" or "Fucking well right." It was everything!

Earlier I found out, when I was giving examinations in the recruiting station in my previous life as a hospital corpsman in the Navy, that this was true of Boston recruits as well—same vocabulary. I had never run into the laboring man the way I ran into them there. And aboard the YMS, most of our guys were working men. This was part of their language. But it just became absolutely mandatory, no breaks. There was not a sentence that didn't have "fuck" in it more than once.

One day at dinner I said, "Pass the fucking mashed potatoes." Then a fit of laughing. And the others did too when they found why I was laughing. How absolutely silly, how incongruous, to talk about "the fucking mashed potatoes." Never have used the word since, even find it hard to say. Long strings of profanity were the most highly savored: "You god-damned, mother-fucking, son-of-a-bitching bastard". This was said to a kink in the cable.

Almost nobody but the Skipper worked out physically at sea. I felt a sluggishness. The sluggishness of eating three good meals and no exercise. You climbed some ladders, sure. But sailors in a mechanical age tend to be slobs.

I was in the glow of mid-twenties, in the peak of fitness. There was no physical thing I thought I was incapable of doing. I needed some of this physical stuff. So I ran. I had a door in my cabin which I could wedge my elbows against and just pumped with my legs up and down just as though I were pounding down the pavement. So I would run for three-five miles (at six minutes a mile). I'd be exhausted at the end of this and dripping wet, grab a quick shower, felt great.

I skipped rope. Now, imagine this on a pitching or rolling or corkscrewing deck. The ship rolls and pretty soon you're swinging the rope at a sharp angle. Then change that exactly the way the roll of the ship went in order to keep from tripping on your nose. I got very good at that. I loved it! The crew grinned but did not join me.

I had another one, doing pushups. If you do pushups as the deck is receding before you, no problem, you can do it with one finger. But you try to do a pushup with the deck coming up at you on a roll.

The galley was the center of recreation for the crew. The cooks would shoo men out at meal time. It was a mess hall with a long table, a recreation hall and cooking galley all in one. There were hatches on each side and there was one by the stove as well. It was a very social place. Always, guys were there writing, playing acey-deucy, poker, having bull sessions. The coffee pot was truly endless. There were good smells during cooking. Then came the disinfectant stink during scrub down. In the dead of night, people would come in for a cup of coffee, bring an arm full of cups to the watch. When it stormed it was terrible; sloshing water, broken crockery, mess. But it was home. It was one place which was home. Day or night. A group or alone. Officers never entered the crew's mess; just maybe to call somebody. It was really out of bounds for us.

We had an adjacent wardroom, two double bunk cabins, a head and a small eating area which had a table: seated four, white linen, silver. There was a glass Silex coffee maker, the coffee machine prototype. This was bolted to the bulkhead so it never spilled and we always kept hot coffee in there. It was good. Glass coffee maker, crystal clear coffee; much better I must say than the stuff in the huge pot in the galley.

It was hot at sea. When we closed up at night blackout, it was terrible. The circulation was very poor. In our staterooms we had one tiny port which you could almost span with one hand and that was the ventilation. We had fans which blew on us. It was a serious hardship in the middle of the night.

I remember one night, getting up in the middle of the night, drenched, soaking wet from sweat, and thirsty. Got up, drank two glasses of water; I had such water loss. In the morning I got up, still soaking wet. I went into the head to urinate and I almost hollered it was so painful. I was urinating yellow syrup, a few spoonfuls. All that water had been pulled out of my body, given off by the skin as sweat and through the lungs of course. So little water was left over that it made almost no urine, in spite of the fact that I drank two glasses of water in the middle of the night. This stuff was so intense, so condensed that it burned. I started laughing.

Of course one part of the tension was the constant threat of action. The general quarters klaxon was right over the wardroom table and when that went off it was unbearable, it was petrifying. We would simply leap into the air when that went off; we had been sitting there playing cards or talking, completely relaxed.

The officers played bridge. I hated it. There were three guys who loved bridge; no, no, let me put that differently, two guys who were good. The other kept saying, "I really must learn to play bridge because it will be good for my career." So three guys wanted to play bridge. Let me tell you, no way, I could avoid learning to play bridge; hated it. They scolded me because they were so damn serious. "How could you not finesse?" "You call that bidding!" Oh my, they were so bitter and I couldn't stand it. I loved cribbage; we played a lot of cribbage. We had bull sessions. We read. We wrote. We censored mail, awful, awful chore. Did our paperwork. Wrote reports. Boring, boring, boring! "Steaming as before on various courses and speeds."

In these pages I have not used names; this does not reflect on my three very good friends in the wardroom. Each was a good officer; I was so lucky! We were very different types, politics, experience, background. And we went different directions after the war; one was headed to banking, another to law, the other to business, and I to science.

We were close, not just in our tiny space but in our duties. None ever shirked. Humor helped; good manners helped; long card games and bull sessions drew us together. They helped me so much; I got advice as well as support. They all followed my lead in dealing with the ship and crew. There was disagreement of course; one of them felt I was too lenient and gave too few courts martial (we only had two deck courts). Our agreed attitudes and policies did not come from formal staff meetings or directives from the Captain. We just got along on the most casual of relationships. Small ship stuff. Why call a staff meeting of four officers who lived in one tiny area?

We were "buddies." This is a term usually reserved for military. More than good friends. We went through it all together. No one flinched. We backed each other up and would have put our lives on the line for the others. That is more than good friends. Such are buddies.

That very major cohesive tradition in the Navy was the Captain's inspection, so important. Saturday morning, very formal. Friday is field day; each department turns to and cleans up to make ready for the inspection. And Saturday I visited every compartment in the ship including something probably most of the crew did not know, the shaft tunnel, to take a look at the shaft that turns the screw, to see that there was not a speck of rust on that steel shaft. Took some doing to get into that. I hoped there were no hidden places, places not seen in the Captain's inspection.

I went into the gunner's shack. And the gunner would say, "Gunnery ready for inspection, sir." And I replied, "Carry on, guns." Then I'd have questions. I would run my finger over everything to find any dust. I am not kidding you, this is a caricature of Navy: I never did use white gloves! But everything was in its place. Efficient. If it's in its place you know where it is. In a storm or in combat, if everything is stowed correctly, you know where to get it without lights. And clean parts work! The whole point of a gun is to fire it; so clean it, oil it.

The engine room was always immaculate! Every pipe color-coded, valves perfect, a sheen of oil. Not a spot of rust. Now you can see the origin of the word, "ship shape."

And there was personal pride. These guys got satisfaction being able to go through an inspection and spend fifteen minutes with me in their compartments and, and I couldn't find anything wrong. Almost invariably I would ask questions; "Did you get those clips for the ammunition?"

"Yes, sir. Got that last time we were in port." There was always some question like this I would remember.

It was a very important task; it meant a good ship and it was cohesive. It was very cohesive. Down in the engine rooms, in the galley, the radio shack. Took the largest part of the morning. And after each division was visited the guys would jump out of their clean clothes, laugh; they were tickled just to get rid of me. Each division craps out immediately, no question about it. Very useful, prophylactic as well as therapeutic. Good for the Captain too, good for the Captain to know the ship and to know the potential problems.

We knew where we were going, the destination. We knew the purpose; to invade an island, to fight, to sweep mines. These were ultimate and dangerous goals. On the way we were escort; slow, boring. Back and forth. "Steaming as before on various courses and speeds."

Drills broke the monotony. We'd time these with the watch. Fire drills. "Fire in the paint locker! This is a drill! Fire in the paint locker!" And the guys would run down there, reel out the hose, spray it. Or we'd time going to General Quarters. General Quarters is so critical. "Bridge, depth charges manned and ready." "Bridge, 3-inch gun manned and ready." These would come up by telephone, or hollered if they were nearby. Every part of the ship recorded the time, cut the time. This was a goal. Drills always carried some resentment but they cemented the sense of mission.

We didn't discuss the global war very often. We largely discussed our own war on a day to day basis. But what did we really do? What does anybody do? An auto mechanic? A truck driver? They just do a job, day after day. And so with us, we just lived and worked on a ship twenty four hours a day. And each day I plotted our estimated position. We were two inches closer on the big chart. "Steaming as before, en route to Hollandia, New Guinea, basic course 198 degrees true, speed 9 knots."

Moored as Before: Life in Port

SEADLER HARBOR, MANUS ISLAND

6. Moored as Before: Life in Port

Did we spend more time at sea than in port? I do not know. It was close. The military is so often described as "Hurry up and wait." We did a lot of that in ports around the Pacific. So this is a companion chapter to "Steaming as Before; Life at Sea." I'd like you to read some sample logs, in their entireties.

> 31 August 1944
> **0000.** Moored as before, starboard side to YMS 360, US Naval Station, Coco Solo, Canal Zone.
>
> 2 October 1944
> **0000.** Moored as before, starboard side to at DE Dock, Pearl Harbor, Territory of Hawaii.
>
> 7 November 1944
> **0000.** Anchored as before, Ulithi Lagoon. Riding out strong winds of typhoon in vicinity, condition 1 of Typhoon Bill set.
>
> 1 January 1945
> **0000.** Anchored as before, San Pedro Bay, Leyte, Philippine Islands.
>
> 18 June 1945
> **0000.** Anchored as before, nearby USS ARG 3, Ulithi Lagoon, undergoing repairs.

I could go on and on. Since those were the only entries made in the log on those dates, this meant we were anchored all day. There was no movement. That was the day. Started at midnight, and that was the day. We had a lot of that. And to any sailor, when you look back, land is truly the human habitat. This is where humans belong. I know, I have said that it truly was good to get to sea, and it was indeed good to get to sea. But when we saw land, it was also good to get off the sea. We changed our minds.

Usually we first "saw" land on radar. Often just mountain tops. Specks on the screen. But they would fit the estimated distance of such mountains. Finally, the definitive shoreline, with points and bays would begin to show up on the radar, and we could see the outline of the coast. We could pick out finally, large buoys, if we were in closer.

But what an event it was to first see the land. The word was passed and almost everybody wanted to be out on deck. There was a lot of talk, "Here we are, what's it like? Are we going to get liberty?" Even when it was only a grey strip, it was wonderful.

I always pored over the charts for hours before a landfall. I wanted to know every part of the landscape, to know the hazards to navigation, what the port would look like, what the facilities were. And I'd memorize the landmarks, their names too. I would know the water depths. I lived with charts. We had hundreds; I don't know, a hundred? We had drawers full of charts. These were in the chart house; it was a tiny room with one whole side a huge table. It was counter high; I think it was green linoleum. These charts were about three by four feet, out flat; you can get a lot of charts in a drawer. All of our navigating instruments were there, and the fathometer, the radio speakers, and a couple of stools. That's all it could hold. I spent so much time there. I think I recall that on the other bulkhead there was an upholstered bench, a place to rest. There were only about two steps up to the bridge, and outside aft was the flag bag and signal halyards.

Coming into port, we often had a man reading and singing out the depths from the fathometer in the chart house. He could be clearly heard on the bridge. That replaces the man in the bow. Remember the stories about "the lead line?" It was a lead weight on a marked line. "By the mark, four-and-a-half, sir." We had one of these, used it as a drill occasionally so that if anything should go wrong with the fathometer we would have guys who knew how to use the lead line. Had a concavity in the bottom of the lead weight; put some tallow on that and then when it came up it was covered with mud, sand, shells; that would give you an idea of the bottom, for anchoring.

Coming into a new port, the charts took shape, out there in the real world. That was such an intellectual transfer. "Aha, that's what this crescent shaped bay really looks like; seems smaller than the chart shows." The islands and bays and points, they all had names. But they only had names on the charts; not on signs out there.

Before sighting shore we were prepared in other ways. With experience in the Pacific, I got so the water told me when we were close to land.

If we were in the lee of an island, the wind dropped or became fitful, and of course the sea changed with the reduced wind fetch. And there was more; the water had more character; it wasn't that clear blue, empty look of sterile open sea. There'd be bits of seaweed, and more fish. Of course there were more nutrients here, more mixing of the land and the sea. Where the nutrients were, you expected to get more biological productivity. And here now were birds feeding and wheeling. After all, even sea birds don't go indefinitely out to sea. They aggregate around islands. And then of course we'd get an occasional land bird, a perching bird, and we'd all exclaim over these who fluttered out this far.

All this time, as we plowed slowly towards land, we could smell it. I don't expect you to believe that. That rich soil smell, the smell of vegetation. In some of those islands we could smell spices and rot, rotten vegetation. And we would see fragments and even rafts of jungle vegetation in the effluent of rivers coming out.

Usually we made port as part of a unit; we almost never were alone. The escorts would make up a line off the coast, an arc of patrol boats, with anti-submarine sonar pointing out to sea while our convoy finally got in to port. Then we would follow them, the last in. Word would finally get down to the individual ships, where to dock, where to anchor, in what order. Often we would go in and anchor in nests on a buoy or docked together on some pier somewhere. Usually we'd be together; we got to know the other YMS's and their people.

Of course at every invasion one of the first things made was the signal tower, the control tower. You always knew, even if it was a very recent invasion, that there'd be a signal tower. They'd have a signalman up there with signal lights. We would send the "dah-dih-dah-dih-dah," and that meant "We want your attention." Or they would "dah-dih-dah, dih-dah" us, and then we would have our questions answered. Often the unit commanders would get the signals and then pass on the various items by our short wave radio, TBS. One way or another we'd get instructions, either by light or by radio, where we should go and what we should do next.

Entering port is always exhilarating, even in a familiar port. Always we had the maneuvers, usually close quarters; docking and anchoring under strange conditions was a testing affair. Always different wind, tide, current. Always, every officer and man is at special sea detail, and kind of leaning forward. The bridge is tense and quiet. And the deck is very busy and noisy getting ready, rousting out the lines and fenders. Or the deck gang is up on the focsle readying the anchor, checking and running the anchor windlass and standing by for "Drop the hook!"

Suddenly the last line is secured or the chock is put on the anchor chain. "Secure from special sea detail. Set the watch." And we'd be landsmen again. This is something so good, so good. The flag would be lowered from the yardarm and as it was lowered another one of the deck gang was raising another flag at the same rate on the stern. And on the bow a sailor clips on the Union Jack, a flag like the corner of the American flag, the blue flag with the stars.

A gangway watch was set on the deck, next to either the gang plank or the sea ladder, and we were ready for shore business. Any visitor salutes the flag on the stern and requests permission to board. And the gangway watch logs this person in and directs him to the officer they're supposed to see or holds them until the Officer of the Watch, is called . . . We were in the business of being in port, and we had a routine for this.

Before I came down from the bridge I had inevitably savored the place. I saw the atoll or the mountains, the reefs or beaches, mangrove shores. And I went over the chart again, reconciling it with reality. We always took bearings on landmarks so that if a storm came up we could check to see if our anchor were dragging. These always had to be true bearings because we were always swinging around on the hook and couldn't rely on something being dead ahead, it might be dead astern in an hour, depending on wind and tide.

The first thing I would do was visit neighbors. We'd get together and have some scuttlebutt and relax and we'd want to know, "Any enemy activity around? Where are the supply officers?" and "Where can we get fuel and water and ammo? Where is the Post Office?"

Everything is so different in port. The heat would descend. Underway there was forced air going through the ship from the movement of the ship. But this all stopped when we were in port. So we had to rely on fans and they never did the trick. The heat would just descend on us and the sweat would pour out.

We shaded the flying bridge with a canvas stretched out to four bamboo poles. It could be removed in a minute. Not very Navy (and not authorized) but we loved it up there, it made coming into port bearable for the watch officer.

The engines stilled. Quiet is not only the absence of sound but the absence of vibration. The sound of the sea, that rushing water boiling by, now was just a trickle. And always that squeal and snapping of the mooring lines or the fenders that rubbed on the side of the pier. Suddenly voices were clear; you heard individual people talking. And of course the port has its own sounds. It may be heavy machinery or ships' horns or whistles or tugs scurrying around putting out big bow waves that started us rolling. Or it may be the desolate quiet of an atoll lagoon where the sea and sky are infinite and the land so trivial.

When the last chore of making port was over, that peace came down. And we let down. One of the first things we all wanted to do was nap. Then wondering what next. When do we leave? For where? What has this place for us? Do they have fresh vegetables, fruit and meat? "God I'd love a steak and salad!" Our cooking always picked up in the quiet waters of port. Cooking at sea is lousy. Tough for the cook to handle big pots of boiling water and frying pans. It's terribly hard. So nice for the galley to sit still.

The ordinary ship's routine does not stop in port. It's just a whole new kind of duty that looms up, and chores need to be done. The first thing I had to decide was, "How can I get ashore?" "Have all the requisitions come to the Executive Officer?" "Where do I go to see the brass?" "Who's the superior here?". I always would need to see him.

Ship's work came first, and what a relief to be able to do it on a level deck. Mail was sought and spare parts piled up (whether from requisitions, barter or stealing). Engineers had the diesels apart all over, each part examined, cleaned and oiled. Electrical panels open, wires and parts repaired. Leaking decks were cleaned and cracks caulked with oakum, a fibrous stuff soaked in creosote, stink. Every man had worked up a list; we had to be "in all respects ready for sea" immediately. Tomorrow? The chaos was deceiving; there was order.

Some chores would involve all hands. For instance when ammo would come aboard or fuel, then up would go flag Baker, the red flag up on the halyards. "The smoking lamp is out;" this archaic Naval command means No Smoking. We'd form human chains and pass the ammunition up into the magazine. This was all hands. Then Baker would come down and we'd go about our business again. "The smoking lamp is lit."

The ports we saw were of three types, with three different feelings about them, three different states of completion. The obvious big, naval base was so different from the forward areas. Places like San Diego, Pearl Harbor or Charleston. These were home bases where everything was available. Second there were forward bases behind the combat areas. In our case, when we pulled into Guam, Ulithi, New Guinea, Manus Islands, Palau Islands. These were forward areas, but not currently in combat. And then, third there were ports being invaded and still under attack; piles of raw supplies, bulldozers, men building, gunfire, tension.

Except for San Diego these were for us all tropical islands. And of course every island is different. The high islands had mountains or hills; usually with a fringing coral reef around them. Usually a barrier reef offshore would enclose a lagoon around the island, and lagoons were very frequently full of coral heads and quite hazardous to navigation. The low island, the atoll, like Ulithi or Eniwetok was a ring of coral reef enclosing a lagoon. The lagoon within this circle of reef, has quiet, clear water, again with many coral heads, some areas worse than others.

I thought Florida was hot, that glaring sunshine. This becomes unbelievable as you near the equator. The heat is just a blow to your body. And the sun is directly overhead. Only place you see a shadow at noon in the tropics is between your feet. You burn to a crisp in an hour. We all tanned, over the months. And you lose water; constant drinking. You just don't move if you don't need to. Everything on deck slowed. And the tar between the deck planks bubbled. Poor appetite. Nap. Sweat; even sweat all night.

At the more forward bases the environment was much more rugged. These were brand new naval bases, built during actual invasions and improved. On low islands the palm tree tops had been blown off by bombing, strafing and shell fire. A city was built. Masses of Quonset huts, corrugated steel. Huge ones. Some end to end, some side by side joined by corridors. Telephone wires on palm tree trunks. Roads, crushed coral rock and sand. The inevitable airport. Steel mesh runway, crushed coral rock. Water towers on rough timbers, pipes on the ground. Hand made street signs. Signs with men's names, signs for a barber shop. Showers, showers of elevated oil drums suspended from a tree or on posts. You get under it and pull on a line which lets water gush. Yankee ingenuity all over.

Then there were the tent villages. Everything open to the sea and sky. Hot, dusty, noisy with trucks running around. Clouds of dust or sand. All over we saw windmills, you'd see these dumb little windmills. And they were home made; one end of the shaft of the windmill was eccentric and attached to a plunger in an oil drum. You filled that with water and soap and throw in your clothes. This windmill would plunge up and down and beat your laundry. Clothes lines were strung between all the palm tree stumps. It was home. It was hated by the guys stuck on the "rock."

The high island bases, on the other hand, were frequently in a valley or at the base of mountains that were forested. This tropical rain forest often went right down to the sea. Frequently the edge of the sea itself was mangrove forest. And there was mud, red mud ankle deep. It was universal; we found this in New Guinea. The valleys on Manus were the same way. The Philippines were the same way.

We paid no attention to the rain. It rained every day, often every hour. You got wet, but you got dry. That's true even in Hawaii. And there was the smell of rot. Mildew smell was overpowering. We took atabrine daily for malaria. And there was "jungle rot," skin fungus, especially on wet feet.

But wherever the Navy went we built as hard as we fought. We invaded and built a harbor; I mean the next day! Temporary piers were run in, hauled on the decks of the LST's; big barge-like boxes that, strung together, made long piers. Then the ships could come alongside and unload. And trucks could go out on these. They would drive pilings, and fill them in with coral debris and have a jetty, a quay. Instant harbor.

The Construction Battalions (SeaBees), could make a city. They had road graders, they had pile drivers. They laid sewer and water pipes, electrical lines. Built sanitation and power plants. Harbors were filled with tankers, supply ships, ammo ships. While the fighting was going on, all this construction was also going on, a naval base in just days. Over a period of time they were smoothed out, they'd try to get some vegetation growing, the junk of battle buried. This was all orchestrated by the local SeaBees and engineering officers. There are many cases of movies on outdoor screens (that was one of the first things built) with Japs watching from the edge of the jungle. They would come in and watch the movies and steal from garbage pails or food supply dumps.

Just a personal comment, being a part of this was quite satisfying. This Yankee ingenuity. Make do, innovate, improvise. It needed patience and hard work but that's somehow what was done. Just for the good of the service. Sounds corny but it was manifest all around and ingrained in our character. Challenges forced creativity and endless work. American.

In port there was so much time waiting for something. We spent many weeks at Ulithi. We spent weeks at Manus, months in the Palau Islands. And always, even though it may have been dangerous, even though there may have been terrible storms, even though there may have been ship and crew problems of all kinds, there was time left over, always waiting for something.

Usually when in a new port, we were quickly resupplied, repaired. All the paper work done. Often at anchor out in the lagoon; deadly quiet. The only routine, ship's work, could be done in a hurry. Then there was, hit the sack. Hit the sack. It was hard to get the crew out in the morning. There would come the banging of a wrench on the steel of the bunks down below. "Rise and shine, rise and shine, swabbies." Moans. And then, "Drop your cock and grab your sock." Another day. We did not usually turn in till late. It was so hot in the evening; when it was ninety, a hundred degrees, you didn't want to lie down and just sweat all night.

We had recreation of sorts, on the ship, and where possible the shore. We read our letters over and over again. We wrote letters. And of course the officers had all this mail censoring to do. We had to read all these things, at least quickly. There was the marvelous innovation of paperback books and pony magazines. This was when they came in use really. I remember *God's Little Acre,* which had many dog-eared pages, dirty lines. But a surprising number of classics were reproduced this way. They were small and very lightweight, and you could get an awful lot into a carton. And then there were the pony magazines. The pony magazines were about a quarter size. And there were no ads; they were great. You could get *Time* magazine, *Life,* and without any ads they were thin. Those made very good reading and I just craved it. The radio news we got was very terse, usually not very good. So *Time* and *Life* magazines were marvelous.

We had a local poet, and the yeoman typed these efforts; there were lots of carbon copies. I remember one that went

A little to
and a little fro
A little ah,
and a little oh.

It was kind of a sweet pornography:

Then was the pillow
from her sweet head
put beneath her hips instead.

Ending:
And wherever I go,
and for how long I may go,
Let me only say,
I hope we will have another lay someday.

This went on and on, about two pages of it.

Handicraft was marvelous. I really got everybody busy collecting shells along the shore and diving for them. Shell beads and earrings were made. We got from the dental officers ashore (I don't know how it was done) a box of silver, kind of like wire. Anyhow, marvelous personal jewelry was made of that stuff.

On Ulithi I came across what the natives called *hani* wood. It was a small tree and its biggest trunks were maybe six inches across. They were bent and wind-distorted. And it was so hard and heavy, I could hardly lift it. And when I sawed it in half, blue smoke came from the saw. I could only do a little bit and the saw got too hot to handle. Then I swept the sawdust over the side when I was through and it sank, it sank like sand! That's how heavy this stuff was. I still have some. I carved a letter opener from that stuff. Very time consuming!

The crew played card games, acey-deucy, craps. There was lots of loose money. What could you spend it on? A little bit at Ships' Services if you could find it. We had beer if I could get it. Terribly, terribly, awful green beer. I remember "Greasy Dick" and "Who de Pole." We weren't supposed to drink it aboard ship, but we didn't have boats to take our men ashore. So we would occasionally close all the doors and put on our blackout lights; we would give the guys two cans of beer, but they had to punch a hole in the bottom so that none of the floating beer cans could be traced down to us!

There were swimming parties organized by the Captain. The Captain had a certain amount of leeway on where we parked so we would cruise in until we found a nice little swimming hole surrounded by reef, coral heads. We would stick our bow in there, making sure we had plenty of room to swing on the chain, then drop the hook where we could go swimming.

I was the first man over the side. We had to put two men with rifles up by the twenty millimeter guns. They could see down in the water, shark lookouts. And if they fired that meant "Recall! Everybody back!" This buck-naked bunch of guys would dive from the deck, or from the focsle. Pearly white bottoms flashing, fuzzy balls bouncing. And hilarity and horseplay, it felt like heaven! This was a delirious wetness! And when there were some coral heads nearby that weren't too deep we could dive down to them and gather shells. They would get rinsed out and cleaned later. We put some knots in ropes and dropped them all along the side; I mean everybody could grab a rope and be aboard in a hurry. The water was over eighty degrees Fahrenheit, really tepid. You could stay in all day, never chill. But it was wet and the breeze so cooling.

Our guys truly enjoyed the chance to frolic in the water but few of them were really good swimmers or thoroughly at home in water. This came to me sharply when our senior engine room man did not want to check our screws by diving for a look.

He had complained to me about a laboring of the main port engine; "I wonder if the screw is fucked up?" When I suggested taking a look he backed off and told me that none of the black gang wanted to do that. He reminded me that we did have diving gear but then told me of his lack of training in the gear. This was strictly non-pressure, hand-pumped air. Remember, only our underwater demolition teams had any knowledge of scuba gear, so new.

This refusal seemed so silly to me. The screw was only down a fathom (fathom: six feet or one armspan of a standard sailor). I did not hesitate; I was at home in the water and knew the gear would work at less than one atmosphere of pressure. So I went over the side knowing I could abandon the gear and surface without getting the bends.

I will admit to initial discomfort. The face mask had a small garden hose from the deck which was fed air by a small pump operated by a man pushing a long pole back and forth. The air was intermittent, hot, and it reeked of hot rubber. I had a chain belt and drifted down easily; the water was incredibly clear, even to the pale seascape of the bottom at 100 feet. And of course, the guy was right; we had some old mooring line jammed around the shaft at the bearing. I surfaced, told the guys and got a big knife. Cutting the rope was tough but I got used to the air. I surfaced in a cloud of bubbles to a grinning crew.

But ashore was the serious liberty, even on small islands. There was organized recreation in some places, horseshoes and maybe a softball diamond. There were movies, there were bars. Beach combing, walking. Manus Island was a place I got to know. We weren't there long when we started out for the Philippines but we turned back because things weren't going as well at Leyte as had been thought, and they weren't quite ready for us. So back into Manus. That gorgeous harbor. Seadler harbor. A lagoon between a barrier reef and island mountains.

In the morning I'd do the paperwork of the ship and go visit offices and do the business. There was a lot of this. At sea you can not do paperwork, you're too weary. And the typist does not work well at sea; the typewriter will not permit it. The yeoman would come down with a half-a-foot high stack of letters, forms and reports for me to sign. Reports all had to be signed. And I'd have the inspections and Captain's Masts. Then we'd chow down at noon. There were no sun shots to take at noon when we were in port.

Everyone would take a nap, and boy, after this I would get ashore. I was able then to play biologist. At Manus, we were on a pier at Lombrum Point, miles from the base but close to the barrier islands. Los Negros Island was very near; that's where one of the main landings was made. I think that's where the airport was. But I could go along the shore and walk for an hour, and be in almost pristine seashores and reefs. The whole island was beautiful, the setting of Margaret Mead's famous book *Coming of Age in New Guinea*. It is very close to New Guinea. Tropical rain forest in the middle and then down along the shore mangrove in places and a lot of beautiful beaches. Palm groves. The islands had been Australian and had many palm plantations; rows of palms. There was the surf on one side of these islands and the lagoon on the other. Barrier islands are much like an atoll island. I spent hours alone here. Outer reef, surf pounding. There was a lifetime of learning to a biologist.

No two islands are alike, but we were in the vast Indo-Pacific fauna. Twenty years later I was to study the reefs in the Indian Ocean. So similar. The same sorts of corals, the same slate pencil sea urchins, the spiny sea urchins, the sea cucumbers, starfish, all sorts of sea anemones and sea fans. And sponge, the sponge in sheets. These sheets of sponge were like trampolines; you could jump on them. And they could be stunning, glowing yellow or scarlet. And the sea would come over the reef and then surge among the low tide rubble of living coral and patches of dead coral. The fish fauna was spectacular, hundreds of species, multi-colored. Moray eels, small sharks; mostly tiny fish in schools.

One day on a Palau Island I was all alone. And I saw something that thrilled me. It would not thrill anybody else! But I saw an acorn worm. This is a primitive chordate, in our own phylum, barely. To me it was a drawing in a text book and it was much larger than I thought, at least finger sized. Astonished! This was really it. This pink crawling thing with that funny head on the end. This was an acorn worm and I had seen it, alive.

Like everyone who has ever been in this region I became a cowrie lover; cowrie is a kind of a snail. They cover their shells with a mantle of living tissue. When disturbed (just touch it) they will pull into their shells and the mantle with it. This mantle gives them an extraordinary glaze, like very fine porcelain. And if you put two or three of them in your hand and roll them around they tinkle like good china. Gorgeous things! The beach was littered with the dead shells of money cowrie, used in inter-island trade. They must be about like a penny, they were so common. But there were the tigers and the gold ringers and other cowries that we gave common names to. There were so many. They could be almost as large as a fist or like a split pea, little teeny ones. I buried them along the shore if they still had flesh and let the insects clean them out for me. I'd come back later and get these nice clean cowries.

Of course it was very apparent to a biologist that it was the coralline algae in the reefs that were so, so important. These were reef builders; they were cementing surfaces. Spectacular algal flora. One of the few flowering plants, vascular plants, to get into the sea was here. This was turtle grass, and some of the lagoons had large beds; they were a world in themselves. There was very little kelp, the kind of kelp that one knows from the tide pools of the Pacific Coast of the United States.

There were giant clams. These were spectacular and all over the Indo-Pacific! They were only visible under water ordinarily. Once in Palau a giant clam had been dredged up on shore; each shell was the size of a bathtub. (I had a face mask, and I also had wooden-framed goggles that I got from Hawaii so I could study these critters underwater). The giant clam, is often depicted in the movies as one that bangs shut on the brave diver who is looking for pearls, and it clamps on a foot. I was surprised that I never saw a giant clam that could snap its shells. They were solidly imbedded in the coral reef. And the fleshy mantle along the edge was brightly colored, I found later, by algae which were symbionts. They made beautiful purple and green edges.

A marvelous realization to a biologist: I was swimming in a soup of biology. It was exciting, I learned so much. I re-read the sections of the two books I brought. I was alone out there. But I was at home out there. I could refresh myself with water from drinking coconuts and eat the sweet white nut. What an advantage I had over everyone else on the ship. The peace of the surf. And this blaze of reef flats, at low tide, and palm trees clacking their branches, and the flowers on the bushes and in the trees. My cup ran over.

It was so pure a world. Yet on the lagoon there were these hundreds of machines of war smoking and rumbling.

One of the first scientific jobs I was assigned when I got to back to school as a graduate student was to write a book review for *Ecology*, the international journal. The editor asked me if I would review a book called, *Fishes and Shells of the Pacific World*. And really, I could bring a lot of experience to it. The idea was to have a review by a guy who'd been out in the Pacific world in the service. It remains my first scientific publication.

I have one overwhelming memory. I was struck by this about a reef. It was all alive! I knew this but now I had actually seen it. The pools of water were streaming with swimming things, large and microscopic. The algae and the corals were all laying down calcium, and this was going to make coral rock; the reef grew. And the surface that I walked on in my sneakers, or swam through, was a solid living carpet or living soup of hundreds of species of living things. I had never truly comprehended this.

In a later chapter I tell of a lengthy stay with the Micronesians on the island of Ulithi, an atoll. At Manus Island I had a glimpse of Melanesians. They were darker in pigmentation and they were taller than the Micronesians. They didn't have the straight black hair of Ulithians, they had curly hair. This is the stock which includes the headhunters of New Guinea, much feared and much debated since then. They differed from Ulithians but I came to know them much less well. I really didn't live with them; I visited them so briefly.

They worked for the Navy in construction; they built clubs and chapels. They did it in local style, a pole shed with thatched roofs and thatched sides and wide open windows. They were very useful and they earned American money, which could be used in Ships' Services, the Navy store ashore. They were able to get peroxide in the Ships' Services; they dyed their hair, and this black hair turned to an orange red, really quite bright orange!

They lived safely and they lived alone on the far end of these islands. They had a fenced off area, a posted area, no encroachment. It was another Navy decision to keep them isolated from the grubby sailors. They were tended to by the Australian military government.

I got to know one of these Aussies because he saw me collecting sea shells on the tip of Los Negros Island. Stunning beach! Unmolested. And I was especially after the rare "cat's eyes" which I found here. These were round snail opercula. The operculum is a lid, a lid to the opening of the snail which the marine snails have but most of the fresh water snails don't have. These were small, not much larger than a bean. And they had a pupil, dark, with white and bright green around it. It really looked like a cat's eye. Stunningly beautiful! Extremely shiny, nacreous. He had a shell collection so he invited me to his home in the village. We strolled about the village. He lived in a very nice thatched house with a wooden floor, off the ground. We sat on the porch and had a beer. We talked of the Manus islanders.

We were interrupted by a woman screaming nearby and he said, "Hey, duck down here and peek through the screen. This happens all the time. You'll want to see this." A beautiful girl about seventeen was screaming at a twelve year old. They were just a few feet apart. And the little girl was laughing. The older girl had a remarkable body; she would have been most welcome in a chorus line anywhere in the world. Extravagant breasts that were conical and (how to put this) they stood rigidly at attention, quivering. The kid was a budding teenager, small body and smaller of bosom. She was teasing the older girl by throwing bits of coral at the twin targets. And the older girl screamed at her and it was obviously gross verbal abuse. Then more pebbles and the little girl would grin and laugh and tease some more. Then finally the restraint of the older girl would crack; driven too far, she started to chase the youngster and almost caught her. But she could only go a few steps before she had to stop and brace herself with her feet apart. The twin targets got out of synchrony; her shoulders were being thrown one way and then another by these things out of their minds. So she finally had to stand braced, waiting for their subsidence. Then another pebble, then some more screaming. It was really very funny. They were so spectacular, these two. My friend said he never wearied of this tirade.

Teasing here, and similar behavior in other islanders, made me ponder the role of this stuff as a more suitable release of aggression or whatever, than physical assault would be. And were not sailors on a small ship also islanders? Our kidding and rough humor—were they the same?

Back to cat's eyes: I gave him my best ones and he was pleased to have them. I was lucky to run into him: it was one of the only places where you could glimpse the local culture. And again, the Navy tried to protect them as they did elsewhere. They had their own village here and there was no sign of trucks or the noise of the construction battalion.

One recreation we had aboard ship was local Navy base radio. We had speakers in the galley and in the chart house up off the bridge. At almost any base, one of the first things that they set up was a radio station, a disc jockey—news station, no ads. The news was official and stuffy and relatively current, usually poorly spoken. Professionally written press releases, all straight forward, official Navy releases.

But I recall sitting in Leyte Gulf in the Philippines off the town of Tacloban, and they had a very good bunch of radio guys in there. They had all this local jargon, something like this:

> Howdy all you G.I.s there ashore and all you swabbies out on the water. This is Radio Tacloban. Let's hear from you with your favorite songs. And here's for you

> guys on that sweep out there on Merrimac 353. Burl Ives with "The Blue Tail Fly" and "Jimmy Crack Corn" and Roy Acuff in "Blood and Glass on the Highway."

Then the guys would all holler and scream, we were included, we were on the radio.

The songs were largely sentimental. I don't know if you have heard the songs of the war-time era, but they had a lot in common and that was sentimentality. "I'll Never Smile Again." "I'll Be Home For Christmas." "I'll Be Seeing You In All The Old Familiar Places." "My Dreams Are Getting Better All The Time." And the old classics from before the war were all there. Swing, the sweet swing bands. "Stardust" was "our song" so it always brought me to tears. There was the bizarre stuff, the "Marezy Doats and Dozy Doats" and "Rum and Coca-Cola" and "Bum Bum Chittum Wa, The Little Fitty Twam Over the Dam," I don't know, but there was a bunch of that kind of jazzy stuff. These songs were heartwarming, though, and they were all melodic, simple tunes, over and over with the old fashioned stuff. But they did the trick. They took us home. They softened the air we breathed.

Sometimes we got Tokyo Rose. We were just leaving Seadler Harbor out of Manus for the Philippines and on came Tokyo Rose; the radioman turned the volume way up so we could all hear it. "You minesweeps, we know where you're going. You won't get there." Response? Mostly laughter.

Almost every port had an outdoor movie with a stage. Movie every night, occasional USO live shows. These had show girls and musicians and comedians; these were luscious show girls and they were scantily clad. They pranced around a little bit and did some dancing. Comedians, unless they were distinctly older, took a bad razzing. Who were these guys? They were 4Fs probably. It was sometimes not so nice and sometimes the audience would get out of hand if the girls were too provocative.

At Manus they welded empty fuel drums together to make a stage and a screen of sheet metal painted white. For the audience, rows of palm logs on the mud. And often at Manus it rained buckets. We had Marine ponchos with hoods, so we made tents around ourselves. We folded them up and sat on them until it rained. Sometimes you couldn't see the screen—driving rain! Show started at sunset because near the equator, where we were so much of that time, night just plunges after sunset. There's essentially no evening; the sun just goes straight down. The bugles would sound "To Colors." The flags all over the base would come down slowly together. Every man on the island stopped what he was about, faced the nearest flag and held a salute until the, "Tut-a-tut-tut-tut-a-tut-a" which meant "Secure from colors." And "Taps" later meant turn in, time to sleep.

At the movie we would stand in lines at full attention, at the call "To Colors," next to our palm logs. At Ulithi they had a ball park on Azor Island. It was there that the "To Colors" at night was particularly poignant because the bleachers overlooked a little cemetery with row on row of simple white crosses. And when the bugle sounded "Taps", it was for them but it was for me and somehow I joined them at that time. But I could look out over the anchorage through the palm fronds. They could not. "Taps" can still tear me apart to this day. It means, end of the day. The clear, sweet notes are stitched into my fabric.

We had a problem with being at anchor and not tied up along side a pier; we did not have transportation to shore. So we improvised, as we always did at everything. Everybody pitched in on this. I would ask our gangway watch to get a boat. He'd go on the focsle and wave his arms at some landing barge or anything going toward shore. One would veer off, pull alongside. They'd stop; lots of joking. They could pick up two of us, one of us, or fifteen. Sure they were going in to the landing, into the boat pool; sure they'd take us right there. "Hop aboard." Friendly. This was the spirit of the times, of the Navy.

At the pier or the quay, when we wanted to come back, we would go up and down and ask all of the ship's boats, "Where are you heading? Are you heading north?" And we'd find somebody who was going back and then we'd frequently load a bunch of crates on. Always someone would be going by our ship. All of us—the small ships—shared this problem.

The officers got to do serious drinking after supper, not too often. I remember so many officers' clubs. Usually they were on the beach, isolated. Like Azor Island on Ulithi, it was beautiful. The place was called "The Black Widow," long bar and many tables. It was a pole shed, thatched roof, no screens. They had iced beer, tinkling glasses, chatter, radio, all drinks two bits. I've never had ten whiskeys at one time since! We would wander in from the beach. This was all Navy help. These guys got this assignment and always there was a steward in charge. This was an official Navy officers' bar. At 2400 the bar would close and we'd get two for the road; we brought them out to the beach. There we would sit on the sand, still warm from the sun, and be right where the water washed up and the palm trees rattled overhead. My God, it dripped with romance! All we missed were the women. But soon we'd had all we could take; had to get back to the pier, back to the boat pool, back to the ship.

I remember one night so clearly. Three of us went ashore. There were twelve drinks a piece. We were stupid with whiskey. But I was the Skipper so I propped them up and leaned them against each other then pushed and coaxed them the half mile down to the pier. I talked to them like a parent for God's sake! Why me? We were equally drunk but I was the "old man," damn it, even if they were older than I. Talked to the crew of an LCVP, it was one of those landing barges that would hold a dozen men easy. They took us out, we lowered these two guys onto the bilge boards; the water sloshed around. They were out cold. I directed the coxswain out into the center of the anchorage. It was in utter darkness, about a mile out. Then I saw them. And I hollered, "353." That means the Captain's coming. The gangway watch was there; he helped me get the men aboard. I put them to bed and then hit the sack for four hours. We were underway at first light, some of us, very tired.

Once in awhile a USO show would arrive. Girls, girls, girls! Or a big hospital ship. Nurses, nurses, nurses! These were . . . these were bad times. We craved them; my God, it was unbearable; they were out of reach. More than unbearable, we were just torn apart by this. We could go for months, we just did fine at sea. We replaced women with fantasies. There were girlie pictures, you know, calendar art. There were letters from home. There were dirty jokes and dirty stories, but we were idling in neutral, we were dulled to women. Adaptive. And then suddenly to have all these gears grinding but the brakes locked.

One day about a dozen USO girls went by, very close, in a boat headed for shore. They were scantily dressed. That was not what we needed. They hollered and they waved and they bounced. I am sure they said, "Let's give the boys a peek; they need it." I was almost physically ill. The crew came unglued; they screamed and swore. There was turmoil. We got no work done. It was devastating, just more than we could take.

One day I was at a ball game over on the island. There were several hundred guys in the bleachers and two very leggy USO girls walked across the outfield. They had on short, short pants and loose halters and they shimmied and smiled. The game stopped. There was silence as they were instantly undressed by hundreds of eyes. Then came the grumble, then the roar. You would not believe it! The stands emptied. All hands poured onto the field. Hundreds of guys hollering, bellowing and running after those two girls. They ran. They ran poorly, but they ran desperately. They got safely to officers' country, where they probably had been invited for dinner. They'd wandered away; foolish women. The story was that only the brass got to these women. They certainly never got to the GI Joe. They certainly never got to an officer of the YMS 353.

The nurses got closer. They were officers in the Navy. In summer uniform, cuter than hell! And sometimes they filled "The Black Widow" over on Azor, and the nearby beach. Two piece swim suits, they'd change in the bushes. It was a mad house! I swear the women acted like ferrets in heat! The male ferrets circled the prey. One senior officer (man) and a gray haired nurse were making out. They were copulating on the beach! They had an audience. It was crummy. It was upsetting. Some of the nurses charged $20.00 and they could get lots in one night. We had a nurse who kind of adopted us. She was neat and nice looking. She'd come to supper. We went swimming with her off the island. That was too close. It was hard on us.

One night at the club there was a shambles. Couples all over on tables, on the deck, on the beach; there was foreplay, there was play, there was afterplay, all around. They couldn't act this way on a hospital ship. It would have been impossible. They had their own places there, and rules. Suddenly, the loudspeaker: "Now hear this, now hear this. All nurses, repeat, all nurses report aboard immediately. Shore liberty canceled, NOW." And boy, did they beat it.

It was an all male society. An all male society can get along. Women, too few and too seldom, devastating. At sea we're together; the internal forces hold us together. But when women, booze, and other things come in, the external forces predominate; in port these alien forces made all of us very vulnerable.

These episodes are still shockingly vivid, because they were so infrequent and upsetting. We saw so few women; almost none of us touched one. In retrospect this isolated, ship-bound, all male population—in a time of war- created a terrible and deep-seated need for booze and sex. I guess we were lucky not to have a crew craving drugs!

In Ulithi one day we were sitting all alone waiting for orders, I think waiting to go alongside a repair ship, a tender. A marvelous idea came to me and it was for my own pleasure, alone. I wanted to rig the boat, the little wherry, which had a mast, sails, lee boards and tiller, and go sailing. The sea was very gentle. Inside the lagoon it never got rough, really badly rough. The wind was fresh and steady as the trade winds are. I had an agreement with the Exec that I was going to head west and then head back east; if anything came up they would come and get me.

It just took moments to rig the boat and I took off due west. I could check with the sun, the wind, the waves, it was afternoon. And I was quartering, the wind was behind me to one side and I flew. In no time I lost sight of everybody, the tallest ships. From sea level you can't see very far. I was alone on the Pacific Ocean, or so it seemed, and it was glorious and we boiled along. Constant fussing with the sheet and the tiller. I went west until I was on the leeward reef. No booming here, this was the down wind side, just exposed reef at low water. I didn't stop and prowl around.

We came about and had one long tack all the way home. The wind and the sun and the waves were on the reciprocal side now. But because I was tacking, I was wet all the way; it was cool and it stung my face. Then I could see the masts of taller ships again and then their hulls and then the YMS 353. Hours later I was in the shower, singing. I felt so rejuvenated and at peace. Oh it was the utmost in pleasure in a far port!

So much of our time was in port and the boring aspect was always present, but if you had any curiosity and some lucky opportunities, you could get around. I had so many special times in so many ports. "I joined the Navy to see the world," and the Navy kindly obliged!

Bullets, Beans, and Washing Machines

REPAIR TENDOR AND BROOD, ALONE ON THE SEA

7. Bullets, Beans, and Washing Machines

As you start this chapter, please hum the stirring notes of "The Stars and Stripes Forever." Picture dawn, thousands of miles west of Pearl Harbor. An enormous invasion armada approaches enemy held shores. In the van, the bombardment fleet plows toward the beach, and above, formations of carrier planes dive bomb that pulverized shore. Salvo after salvo of continuous thunder. How many rounds of ammunition large and small? It is awesome, theatrically splendid and does indeed call for those martial notes you have been supplying.

Yet, looking back on that devastating horror, the astonishing thing was the supply system that put all those shells and bombs on all those hundreds of ships which came from bases all over the Pacific to form this fleet, at this place, at this time. Our YMS was a pebble among those boulders.

Just one of those battleships alone had a two-three thousand man complement, and the men on cruisers and destroyers, all those transports, LST's, all those troops, all had to be fed. At Four o'clock in the morning there was a chow call throughout that fleet. They had this incredible breakfast! Eggs. How many eggs does it take for one crew of 2,000? How many crates of eggs for one battleship? How many tons of bacon? They surely had bacon and eggs with toast. And how many hundreds of barrels of coffee just for one ship? Where did they keep this stuff? After a month at sea? This was incredible!

By comparison, supply was nothing on a ship with thirty-five guys like the YMS's, nothing. In the class of YMS, the Y stands for Yard, Navy Yard, MS for Motor Sweeper. The "Y" meant we were to operate out of a Navy Yard, like at Norfolk Navy Yard where I operated for a year. We went out and swept the approaches to Chesapeake Bay off the Capes there. A hundred miles we went, a hundred miles back, a hundred miles out, a hundred miles back. We would do this for five days. Then we would be relieved by another team going out and we would go in. The YMS had a five day supply of water, fuel and food; that fits the YMS designation.

After five days we tied up securely, engines quiet, everybody relaxed; we rocked very gently on the pier there, lines creaking and groaning. But before we settled down, the gunner, the cook, the chief engineer, the pharmacist's mate came down to the Executive Officer with a long list of supplies that were needed for the next five days. I signed these routinely. The men knew what they needed. And then they would go off whistling down the pier and find their engineer shack or food supply warehouse. Medical supply shack. Quonset huts all over the place.

They returned and the trucks would start coming. A truck would come with one great engine part or a spare reel of cable, minesweeping cable, or anchor chain that we had to have replaced; one truck after another. The pharmacist's mate would just get a small cardboard carton but it would have a whole bunch of syrettes, bandages, first aid stuff, medicine, routine medicine. The yeoman had paper, pencils, forms to fill out.

The cook ordered two sides of frozen beef; I don't know how many crates of eggs he would get at a time. Five days, thirty five guys—a lot of eggs! Coffee, gallons and gallons and gallons. But every five days we were resupplied. This was the way our ship was intended to be, supplied for five days. That was deemed all that we would ever need even if we had to go a couple hundred miles up or down the coast, we could get back in five days. Unfortunately, that was not always the case, as you well know; we took off for far places.

Pacific distances magnified problems enormously and I still am amazed at how many ships were supplied all over the world, on all the seas and oceans. There were 800 minesweeps alone, 400 of those were YMS's (In the '80s we heard a President saying he wanted to restore the Navy to 600 ships). There were thousands of ships in WWII. All over the world. They got their mail. They got their fuel and water when they needed it. Huge supply ships in convoys were headed to all the places where the fleets would be. They refueled at sea; many had their own water stills of course. But all over the world this was going on. We were such a tiny part of it.

The problems for a YMS came from being so scattered over the world. Who could keep track of 400 YMS's, much less their needs? And, as I mentioned, the ship was not designed to be far from a base; it was not authorized to have the gear that fleet ships had. And our mission changed; we were not designed to do battle with planes. We did though. Some of this the Navy could not help, could not have foreseen. There were all kinds of problems with any such system—smooth as it worked, incredibly smooth. I want to tell you a bunch of those stories.

One of the things we took for granted whenever we worked out of a Naval Base was laundry. The men all got together and each section, black gang down in the engine room, the deck gang, the bridge gang, the cooks and the wardroom. The officer's steward took our bags of dirty clothes. Next day the truck came down with our laundry and it was gorgeous. Shirts were starched, pressed, everything was neat and nice. I have never before or since had laundry so good.

But when we took the trip from Charleston, around through the Canal via Miami, via Cuba, came up the west coast of Central America and Mexico, we had problems. After about three weeks we got in to San Diego, where we were to be for a week or so for our last crack at outfitting. I immediately went to the Base Commandant, along with the other skippers, in our little fleet of YMS's coming around. We paid our respects, and the Admiral at the San Diego Naval Base said, "Gentlemen, whatever we can get you, you just ask." And of course the first thing that occurred to me was a washing machine. By that time they had automatic machines; the Bendix was out. It would be marvelous to have a washing machine out on a deck and then each section could have a turn.

After three weeks of coming around through tropical America we had absolutely filthy clothes. We couldn't get them cleaned. We could only use salt water and salt water soap, then scrub, scrub, scrub on the wooden deck with a brush. Then drag it over the side to rinse it in salt water. It was just a mess. Clothes were terrible looking and feeling. I kept one starched khaki suit to wear on the visit to the Base Commandant.

We went immediately down and made application for a washing machine to go on the deck right behind the officer's wardroom. We had a potato locker behind the wardroom; it was a giant thing with slatted sides. We didn't use it for potatoes; It was not where you'd want to put potatoes, the tropical sun beating on it. We figured that spot would be right above the engine room and we could pipe it with hot water, fresh water. But we weren't allowed. They looked us up on the authorization chart. A YMS was not authorized to have a washing machine. (Do appreciate: the USN worried about weight. But at seven feet above water?) So much for that. After all we only left port for five days!

So much also for my request for an evaporator for fresh water. We almost ran out of fresh water a couple of times, even though we were only gone a few days. But we were not authorized for an evaporator. After all, we only left port for five days.

We had the most marvelous officer aboard who was a businessman in civilian life. Older than I by a few years; he was the most genial, perfect salesman. He smiled, great grin, and said, "Skipper, I think I can get us a washing machine, because I happened to notice there were hundreds in a warehouse down there. In fact, there were so many washing machines that they were even stacked on the pier, out on the concrete. Without a tarpaulin over the top, in crates." He said, "I can't imagine that I can't liberate one of those."

I said, "Well if you do, don't let me know about it."

"OK." He got about three guys from the engine room and they went down to the motor pool and found a truck with the keys in it. They borrowed it; they got a truck. They had written out a requisition for a washing machine. Just wrote it out. There were all kinds of forged signatures on it. They drove down to the pier (this was all hearsay, I wasn't there); they backed up to this huge stack of washing machines. They were all going to New Guinea; all these washing machines going to some big naval bases, army bases, airports, whatever. And there was a Marine guard, a sentry, walking back and forth. This was in the evening, night in fact. And they backed the truck up there, and our officer just called his men and waved his sheet of papers and said, "You got to look for the right number now." He had written down a number of one right near the edge, "There it is, there it is! Read off that number."

He read it. "That's right." All this within earshot of the Marine sentry who really paid no attention; this probably was going on all day. They pushed and pulled. It was gigantic. Anything in a crate looks ten times larger than it is. They had some planks; they got it up on the truck. These bandits just wiped their hands of this whole thing, got in the cab and drove back to the ship. We immediately had a winch running on deck, a topping lift for lifting an awful lot of weight. They lowered that boom then up it went. The potato locker was gone. The engineers were there, they cradled the crate down. The gunner's mate had made a canvas hood, a formed hood of canvas that would just fit over the potato locker and painted it gray, battleship gray. They immediately dropped that over the whole thing.

When we finally set out to sea they installed this marvel. They had hot water, everything right there. And within days of departure of San Diego we had a schedule of laundry for this machine. You could just put the stuff in, press the button, walk away and leave it. It would spin dry. We had a day for the deck gang and a day for the engineers, a day for the steward, a day for the cooks and a day for the bridge gang. That meant that it was busy every day of the week; we had clotheslines strung around. Needless to say that was the end of the shirt ironing and the starch, however. But they were clean and they were fresh. And when you put them on a hangar and hung them in the wardroom they hung out quite nicely.

There was a way of getting around an unauthorized item, but not legally. The Navy was silly; they could change our mission but they could not authorize a washing machine. We won!

I mentioned a salt water still for fresh water, an evaporator. I talked with our top guy in the engineering crew. Extraordinarily competent mechanic, steam fitter, diesel mechanic. And several years in the Navy, an old hand. I asked him, "If we could get some parts do you think that we could make a still?."

"Oh sure, oh yeah. The idea of a still is very simple," he told me.

I said, "I know it is, but I wouldn't know how to start. We can't get one, not authorized." I said, "Do you suppose (and now I must admit to being somewhat Machiavellian here) without disobeying any orders, without stealing anything outright (like we'd already done), how about requisitioning spare parts, as though we had a water still?"

"Great idea."

"What shall we start with?"

"Copper tubing." He said, "If I can get copper tubing I can make, almost make, a still out of stuff we have aboard."

We did that in San Diego. We got copper tubing. A whole cardboard carton of it all coiled up. Plenty of tubing. It was replacement, spare parts. And it wasn't a still. Subsequently at Pearl Harbor he picked up the rest of what we needed to make a still, a piece here and a piece there, a pump here and a brass fitting there.

They had that thing all hooked up and ready to run before we left Pearl. What a relief! Now we not only had clean laundry but if we were really caught, if we sprung a leak in our fresh water tanks or if it were holed by a shell and we lost our fresh water, we could make enough to get by. We could make enough to cook and enough to at least keep clean and to drink during the hot days that we knew were coming. This was an example of getting around the Navy regulations, bending them but not really breaking them. The Navy was silly. We won! Again!

The biggest scam that the YMS ever pulled off was done in Pearl Harbor. I again knew nothing about it. As a matter of fact I truly didn't. We were tied up alongside a pier near a machine shop, having some work done in there. We'd been in Pearl three or four days; in fact I think it was just the day before we were going to leave, or maybe it was the last day. I heard this commotion, couple of guys, loud voices. Then feet pounding, running around on the deck. There weren't very many secrets that way. They were right outside my cabin port. I jumped; something had gone wrong; and I headed out. A man was standing outside of the entry to the wardroom. He said, "Skipper, I don't think you want to see this."

"Listen, what's wrong?"

"It's nothing serious." And then he grinned at me; this was a man I trusted a great deal. He, he just grinned, and said, "Just, leave it to me." Then, "Wait a minute and I'll tell you all about it." He disappeared. I was absolutely perplexed. The running stopped; all was very quiet. I sneaked a look out. Nobody in sight, up and down; it was just quiet. Guys working on the deck, just doing some painting, chipping. People walking around doing their jobs. Then my informant came in; he was a first class gunner I think. And he said, "I got a present for you; can't show it to you, but I got a present for you."

"What's this all about? Is this what the fuss was on the deck?"

"Yeah. Here's what happened. You know, up by the administration buildings, before you go out the gate, there is the longest row of bicycles you ever saw. And the thought occurred to me, we're going to be going out on all those islands and the skipper doesn't have a jeep, he doesn't have any kind of a vehicle. Wouldn't it be nice if he had a bike. So I found one that said 'Commander so-and-so.' I don't know who it was. But you know with that kind of rank he could get another bike. And I just got on and biked down here. Nobody said a thing. After all, guys are biking all over." And he said, "The engineers were ready, and as soon as I got to the edge of the pier with the bike, and the tide was out, we were down a little bit." He said "Threw it aboard and two engineers caught it and took it to the far side where it couldn't be seen from the machine shops or the pier." I pieced the story. Two engineers went to work and the wheels went immediately into the engine room. And then the frame. In the engine room they took it down to it's last bolt. And they distributed these parts all over the ship. In the mess hall, in the galley. They left some down in the engine room. In the various lockers of various kinds. Paint storage—there were the handle bars.

After he told me about this (I couldn't help but chuckle) I watched them, after we got to sea. They were like a bunch of little kids. They knew where all the parts were and they assembled them all. These guys had put bikes together, had grown up with bikes. They immediately wire-brushed the bike with a rotary electric steel brush, and took it clean down to the raw steel. Painted it zinc chromate yellow, anti-rust paint. Then finished it with a nice blue and gray. Lovely thing. And they got a new plate and put "YMS" on it and the number. They had filed off the identification marks that were there. The stamp, "USN" was left on.

It was hauled out, all shiny, and I was there to see it. Everybody wanted to make a presentation of this bike to me; and I couldn't help but laugh. As a matter of fact, I was tickled to death at the thought. And it was useful, it was useful. But that was grand larceny. Never heard a thing of it. I'm sure it was published in a newsletter around the base, "Anybody knowing of the whereabouts etc, etc, will be severely punished, etc, etc, etc." And we never heard a thing about it. Thank heavens. It was useful and it was an awfully nice gesture. The Navy did not say no to a bike. Never occurred to them, and we never asked for a bike. I must admit it was grand larceny on our heads.

In October, 1944 I attended a big meeting at Pearl Harbor, Headquarters, where a whole bunch of skippers were assembled. We had all the details from the Flag Lieutenant of the convoy. The Commodore was a Captain, a Reserve Captain, former merchant marine skipper I believe. He was on a large freighter as flagship. Basically this was a flotilla of LST's going to New Guinea and we were one of the anti-submarine escorts. There was nothing but small craft escorts. We were going to New Guinea via Eniwetok in the Marshalls, Guam Island in the Marianas, Ulithi in the Carolines, and then down to Humboldt Bay in New Guinea, Hollandia Town. Different name now.

First stop after we left Pearl was Eniwetok. And we knew that from then on we were in forward areas. Eniwetok was one of our first atoll invasions. And it showed the scars of war, which we saw for the first time. There were no palm trees anywhere, just tall stumps. And this mass of Quonset huts and tents; this was a huge base. It very quickly became a very, very rear area, especially as Ulithi, where we were headed, was just now growing as a forward supply area.

We got to Guam and by that time we were on a full-time war alert, dawn and sunset General Quarters, although the Marianas had been secured in the summer and this was two months later. There was still sniper activity, and we were urged when we got there, by our commodore, that we were to show no lights whatsoever. From time to time uncaptured Japanese were sniping at anybody who was silhouetted against a light out in Apra Bay.

We got to Guam in the night so we made a run, I don't know why, clear around the Marianas. We went clear up around, up around Saipan, Tinian and Rota, then back down the west side of Guam and came into Apra Harbor at dawn. But it was a gorgeous sight. There was a barrier reef to our left and on our right was the Orote peninsula with the airfield on top. It was here that occasionally snipers would appear. It was a high bluff up there, a fringing reef below. The active harbor was only about, one by two miles. There was a lot of reef, very foul ground at the far eastern end, tidal reef, exposed coral flats at low tide.

The harbor was filled with shipping. When we came in you could hardly move. One of the first things we did was ask "Sara" (our commodore) what we should do about provisions, fuel and water and ammunition. He said to contact the local commander, whom we somehow found on TBS (short wave radio); there was just a chatterbox of stuff going on all over. People calling. They said "Commanding officers use own discretion." Fuel barges were down near Cabras Island which is northeast; it was a very tortuous way down there. And a whole bunch of ships were waiting to fuel up. We were down to one-third fuel; we had fueled at Eniwetok and once in between, at sea. So we talked with some of the other YMS skippers and they had found out here and there that the thing to do was to get hold of some ships that had just come in from Pearl and see what they had.

So much for the miraculous network of care for the naval vessels! If you were a cruiser in a battle fleet, somewhere you had a rendezvous with everything you needed. At a given time and place. But if you were a YMS, escorting a dumb little flotilla of LST's, average forward speed 10 knots without zig-zags, that was not something any one had any notion of.

My marvelous officer, the scrounger, said, "You know, that LST that we were escorting, I talked during the trip (you know on the light, on the blinker) with an ensign over there who I went to school with. Fraternity brother. I'm going to give him a ring." So he got a hold of him on light, and asked the basic questions, "Do you have diesel fuel? Good water? Bread? Meat?"

There was a little wait and he came back, "Sure. Come alongside." So we upped the hook and wandered through the maze of ships, found this LST and pulled up alongside, amid all kinds of good revelry between crews. We lowered some big fenders; we rolled in the gentle swell. It was a very open harbor; so there was a good swell.

We were able to pick up what we needed, almost everything we needed, from this LST. They took our hoses over, we got fresh water, we got fuel from their spare tanks. They had bread. We got bags and bags of bread, and we got flour. So we had fuel, water, flour, and bread; we knew we didn't have a long run till Ulithi which would be our next stop where, presumably, there were more supply ships.

But that was certainly our very first taste of food scrounging. This was the high art which we had to develop. The cruisers could get supplied, resupplied. All they had to do was to be at a given place at a given time and the resupply train would be there for them. Not so the spit kits, the little things, not so all of these uncountable, unnamed, small ships going all over the world in need of frequent and constant resupply. We had not decent storage space for anything. Five days! A helluva mess on a YMS.

But here in Guam we got everything we needed; we were patting ourselves on the back and we kept hearing from these other commanding officers by light and by TBS, "Did you find any fuel, did you find any fuel? We can't get fuel, we can't get fuel. We can't leave without fuel." We were indeed resupplied and it was the first of many, many such scrounging efforts all around the Pacific. We were initiated to the forward area.

"The Skipper is a genius;" the crew knew this after the famous flour beetle trials. We got our flour in sacks; huge, heavy paper sacks, olive drab sacks, flour. Wherever in the world it came from, I don't know. Anyway, we had it and the cooks were very good. They knew how to bake bread in our ovens; they had control of it. They baked gorgeous bread; they made beautiful toast. One thing we had was bread!

When we were anywhere near supply ships, we could get butter. When we weren't, and this happened an awful lot of the time, we got what they called "duck grease." This was the first attempt at synthetic butter made of margarine materials. But it was almost liquid and, and you sort of smeared it on toast and held your nose while you ate it; it was pretty bad. But somehow it was yellow and greasy and a little bit salty. This duck grease was awful. We did have cans of jam for toast.

But the flour got full of weevils, a kind of beetle. Now, these I knew. When I was an undergraduate at Chicago, I worked for a year in a flour beetle laboratory, counting them, sifting them out with sieves so we could count all the different larval stages with different mesh silk screens. We did the demography of flour beetles in competition experiments. So flour beetles, care and feeding of, handling of, I knew.

The cook came to me, worried. I had noticed these beetle parts in the bread, whole wings and half bodies and legs. One tiny weevil was the size of an ant. But the fellows didn't like it. They told the cook to get some flour without it. He'd open another sack and in no time there'd be beetles in there. I told him they weren't harmful but that didn't help. Finally I said, "Hey, Cookie, see that brass screen in my port hole, you take that and run all your flour though that screen into a huge box. You will not have any beetles." Only I, I mean only I, knew that 90% of the weevils were still in the flour, namely all the grubs, all the larvae which were still in the flour but they were white and just cooked out when it was baked. It was protein reinforced bread! And everybody was thrilled and I was quite a hero. I knew how to get rid of flour beetles: you run the flour through a screen.

"You know, those, those sacks of flour for some reason," the cook said, "kept getting more beetles." Of course they did. All the larvae, all the eggs were going right on through and then within weeks there was another batch of black adult beetles again. However, there was a continual screening, they'd come and get the screen and run all the flour through and in a couple of weeks later, he'd be back, "Skipper can I borrow your screen again?" Everybody was happy.

I went ashore at Guam to check on mail. There were some dispatchs for the ships, nothing of any consequence. But no general mail and everybody was bitterly disappointed. But heavens, we'd only been a couple of weeks out of Pearl. We had gotten mail in Pearl, big sacks of it. Fleet P.O. knew where we were headed. I promised the crew we'd have some in Ulithi which was our next stop. Straight south of Guam, a thousand miles or so, and our convoy was heading that way.

We got to Ulithi very quickly, seemed to us. We had our first red alert on the way. That is, a Japanese plane was sighted and we all went to General Quarters late in the afternoon. But we were not attacked. Nevertheless it was very sobering. And we, all of us, felt that now we were veterans, we'd had our first "action." Great! We just saw this plane in the distance and the flagship with a bigger radar, aircraft radar, had tracked the plane. This was a red alert, a real one, but what a gentle introduction. Nevertheless we were sobered and a bit more salty.

When we got to Ulithi we were absolutely flabbergasted. This was our second atoll (we had, after all, come on Eniwetok from the sea, so we knew about atolls) and we saw the usual cloud formation over the atoll. Ordinarily there's enough vegetation to generate moisture, enough for clouds over an atoll. This was one of the ways that Polynesians and Micronesians found their way all over. We then could see the tops of the palm trees. Nothing higher than the tops of palm trees. The island was nowhere more than a meter or so high. It was awfully exciting and the convoy limped in there at our ten knots. We were met at the reef entrance by a sub chaser that was going back and forth across the entrance, two miles this way, five miles that way, ten miles this way, guarding the entrance. The atoll had a very narrow entrance, only a few hundred yards wide between two reefs like shark jaws. A pass so narrow, but deep, deep water and we sailed right through there single file.

We found our designated anchorage and dropped the hook. This time we were no where near a land mass. We were in no way at a Navy yard. But the astonishing thing was, they had made Ulithi into a floating Navy yard. There were floating dry docks. There were huge auxiliary ships, battleship gray, but they were very large merchant ships converted to machine shops. And you could go alongside of those and get availability, get your engines replaced. They could do anything.

There was a long string of huge barges, the length of a football field. Concrete barges, hollow. It was a big hold full of supplies, and on the deck they had some booms for cargo handling and a mast for the radio. They had a deck house for the crew, a very small crew, just for cargo handling. The barges were anchored fore and aft. Huge, huge anchors, they couldn't be moved.

Ships came alongside and left an order; then they pulled off like a drive-in hamburger place, you know. The barge would blink back; "We have your order in." And you'd wait your turn. A line formed to come alongside. They had a cargo net filled with cardboard cartons, down it went onto your deck. And then, thumbs up, back went the net and the cargo boom and you'd pull away. Each barge carried a specialty. On one barge they had nothing but machine parts. One was for mail; we went there and we had mail, two sacks of mail. Incredible that they knew we were going to be at Ulithi at this time and would stop and get our mail. General Delivery, Ulithi.

This was an atoll, it was twenty miles long, half that across or bigger. And it was purely a ring of reef and very shallow water, not a hundred feet deep in the middle, and most of it less. Anchor anywhere. The reef was punctuated by little beads of land, islands. And there was one called Azor built into a base in the northeast corner. There was a small airstrip on nearby Falalop, I believe, small bombers, fighters, small cargo planes could come in and out. Azor was covered with Quonset huts. There was a baseball diamond with a backstop and bleachers. They had an Officers' Club, "The Black Widow." There were churches and chapels, a cemetery, and movies on Azor Island.

Another island, Mog Mog, was a purely for recreation. The ships would come in; lower their barges over the side with half the crew, go to Mog Mog and they'd get soused with beer, pitch horseshoes, play softball and come back sunburned and stinking and exhausted. But they'd been ashore, maybe after a month or two at sea. And it was wonderful to have the shore there. Tiny bits of island on a huge ring of reef. Ships were entirely in the top one-quarter of the lagoon, inside the reef, the anchorage.

And one time I saw a dozen huge aircraft carriers there. I'd see three or four battleships. An endless number of LST's and LSM's, LCI's; landing craft of all kinds. And cargo ships, and cargo ships. Two or three hospital ships. Everybody was in there, anchored. It was the biggest anchorage in the Pacific Ocean, and it held more ships than any place I had ever been. Everything was afloat; you went from one to the other and got your supplies like a shopping district. It was really pretty marvelous. One, mind you, did nothing but make Coca-Cola. Another one stored cold beer where you could go and get your quota of beer. Another one was an ice cream barge. Get several tubs of ice cream, if you had room in your freezer for it. It was staggering, characteristically American. Coddled American servicemen in the far Pacific. The Japanese would have just been disdainful, it was so un-Samurai of us to drink Coca-Colas at night and have a dish of ice cream in the middle of the war!

It was button down at night; it was absolute black out integrity. We were just a ghost at night; you couldn't see a thing. It was really quite grim. Nothing moved on the islands, no open lights at all. And there were all kinds of people there, walking around, in the dark. And there were boats running around from ship to ship, but completely in the dark, without running lights; you had to know where you were going.

After our first scare with an air attack we felt that we were really underarmed; all hands agreed. We had a three inch .50 caliber, the seventy-five millimeter bang on the bow. Just aft of each wing of the bridge we had a single twenty millimeter on a gun platform. And that was a very powerful machine gun. Then on the fantail two twin .50 caliber machine guns, a pair back there on each side by the depth charges. It really didn't seem like very much anti-aircraft fire. We simply were never designed to operate where we would have to shoot airplanes.

We returned to banditry at our next port down in New Guinea. The boys were "cruising" and they found an enormous stockpile of guns on the edge of the jungle. Out in the open, under sheds: hundreds of machine guns, larger stuff, all kinds of ammunition. Ships were coming in and they were getting AllNav orders, "You are to get rid of all your .50 caliber guns and replace with twenty millimeter." Or "You are to get rid of your twenty millimeters and replace with forty millimeters." Everything was being upgraded, and so ship after ship would come into New Guinea and drop all the .50 calibers into a barge and get twenties, and go out proudly. They stashed the discarded stuff over in the jungle. Our guys found this stack of discarded .50s back there.

We had another "midnight requisition." Our guys went in and chatted with the man who was in charge of this stuff. He agreed "as hows it really wouldn't matter if anything were taken;" but he couldn't authorize it. And as long as they were talking he was not looking and half our crew was over there hauling off .50 caliber machine guns and box after box of .50 caliber ammo. They lugged six single .50 caliber machine guns back to the ship; half the guys staggered under boxes of ammo.

Back on the ship the engineers just loved this stuff. They had the same zeal they had for putting in the washing machine. Along the gun'l (along the rail) of the main deck, they put three on each side. They were welded onto plates, steel plates, which were bolted into the deck, with braces to the gun'ls, so they were really thoroughly, thoroughly anchored. Over against the engine room housing we mounted these ready ammunition boxes, which were water proof with dogs and rubber gaskets; so they were right out on deck by the guns. And at General Quarters they could be undogged in a hurry by spinning the spanners, and, and there you had all these clips of .50 caliber to pop right in the gun.

That meant we were a porcupine of machine guns. Now almost everybody, including the cooks, was on some gun crew or an ammunition handler. From then on, in action, we put up a curtain of lead from six extra guns, a big help. We could put up a terrific broadside of machine gun fire with those things! Another unplanned venture, an unauthorized mounting of guns contrary to Bureau of Ordinance authorization for the YMS. We won again.

We were assigned to Manus Island in the Admiralty Islands from nearby New Guinea. Manus was a well established base by now. We had there, ashore, all the naval base facilities, including laundry. We really didn't need laundry at this time, but we did take advantage of the starching and ironing when we could. We had requisitions, and we got all the food; we got resupplied with all the ammo. We got spare parts for everything including some new coils for our evaporators (we kept getting spare parts for our evaporators even though we weren't authorized to have the evaporator).

I want to go on to one other tale, one other tale of skullduggery, here in Manus. We had the boat (I've mentioned it before), the wherry. The little row boat was 12 feet by 4 feet. It was terribly useful and we were very glad to have it. But the drawback, it was little. It was tiny, held two people, one rowing, one sitting in the stern. No more. And if you were in a place like Mayport, Florida or in a nice little harbor like Hollandia, with no big winds, you could row it around. But you couldn't go to shore and load anything in it. If you went to shore to get mail and you got two sacks of mail, it would sink the wherry. It was no good that way. And I was unfortunately the only guy who could row it well, no one else really wanted to, I think.

It occurred to me though that it would be wonderful if we could get a small outboard for this thing. Almost anything would help. Then anybody could buzz around. We could increase the efficiency of this going back and forth from ship to ship in the wherry, which was the only real use for it. So I made that long trip in at Manus, into the base. Got a ride in with our usual technique. Thumb, holler: "Going in to the base"?

"Yeah." And you'd tell them it was Fleet Landing, Lorengau. They had big piers coming out, and it was full of boats, so you could always get a ride in. I found my way down to the supply officer and told him my sad story. That I was a veteran of YMS duty on the Atlantic and the Pacific, and I told them places we'd been in the Pacific. I said, "I'm a veteran, I know YMS's and I know this wherry and I know that it has this terrible limitation." And I said, "I know BuShips hasn't authorized an outboard for the wherry, but don't you have some outboards we could use."

"Oh, yes, we have outboards. I just hesitate to do this."

"Well, this once. We're about to head into Leyte Gulf." I said, "We'll be going there next and we'll be running around in the wherry, I know we will."

"OK." So he called over the yeoman who sat down and rapped out the requisition, "Outboard, Mark two, Mod one." I don't remember, I don't even remember what brand it was. Maybe it didn't even have a brand name on it. And then he wrote in under "Amount," the number 1. I got it, put it in my pocket. Felt great. And I asked, "Hey, could I get a truck? I want to take it down to the jetty."

"Sure, sure, sure. They got one up there. Tell them you need a truck."

I found the Quonset hut where these things were stored. Ha, what a place! Everything stacked up to the ceiling. They had one. Then I looked at the requisition. The "1" written on the requisition was curved, kind of a crescent-shaped "(". I said, "Oh, my God. With no trouble at all I could turn that '(' into a 6." And so I went into the office there, and while I was waiting for this guy to come who knew about outboards, I just picked up a pencil and I changed that to a six. It was undetectable! I smudged it a little bit; it was nice. So he took it, and said, "Six outboards."

"Yeah, yeah," I replied. He came back with them. Oh my God; what a mess of outboards. Huge crates, heavy crates. Got them on the truck, and down on the pier. Went up and down looking for somebody going out past Lombrum Point, "Got a load of crates here." Finally found somebody, a large landing barge.

"Sure, I'll take you out." The coxswain was genial. So he took us out. Well, I was Santa Claus. Went from one YMS to the other. Shouts of glee. I got rid of five, had one left. I came over to the 353 at last. Hollered up, "Hey, here come's an outboard." Ohh, the motor machinist mates were so tickled. They loved outboards. Already the other ships were running around in theirs. They came out of the crates fast. We got ours out of the crate in a hurry. Put it in the water. "Hey, look at this guy go!" Wouldn't start. It wouldn't start! Then it would start; it would die. Wouldn't you know. Six outboards, one lemon. And we got it! I had forged another victory over BuShips authorization. But we lost this time.

Scrounging for supplies was a shock to us in Guam. It became a way of life. The problems of supply increased as we approached the combat areas. Even in rear areas like Guam and Ulithi and New Guinea, there were some problems; scrounging became acute, but there were things around if you could find them. When we really got in the combat zone we faced privation. It was serious. Can't you see how that could be?

Food was never rationed, but it was poor. It was very poor. Meat meant Spam in a can, or baloney sausage, affectionately known as "horse cock." They came in about three foot lengths and three inch diameters. We ate powdered potatoes, powdered milk, dried soup, powdered eggs (which when they were cooked into "scrambled eggs" had a strong ammonia odor). How we ached for a chop or a steak, stew, or even hash! We'd talk about it, remembering. But we also remembered seeing troops ashore eat cold Spam in the jungle rain and mud. The YMS was comparative luxury.

Even more we longed for fresh vegetables and fruit, salads. This was the debut of powdered and frozen foods, when Bird's Eye got its start. This was a big help. But can you imagine being without fruit in the tropics? There were coconuts and banana trees all around us. I ate vitamin pills daily; probably helped me a great deal. But I'm quite sure my teeth suffered a great deal at that time.

One of the rotten days I remember was in Leyte Gulf. A supply ship about a half a mile away started throwing crates over the side. What in the world were they doing? Then some of the crates drifted toward us, full of what looked like cabbages. I immediately jumped in the wherry and got some guy with me, and we went out and looked at it. Sure enough it was cabbage! They were sort of rotten. This ship was dumping over the side all of their vegetables that were a little bit tainted perhaps. And so we gathered up potatoes and cabbages, and carrots; here they all were bobbing in the water. We immediately picked over the rottenest ones and threw them over the side. We filled the wherry; with vegetables, bushels. Potatoes, cabbages and carrots and we took them back to the ship. There the cooks carefully washed them and cleaned them. We just grossed on all these fresh, crunchy vegetables. It had been so long.

One of the items that we could get from a concrete barge in Ulithi at our last stop was candy. The one labeled "Ship's Stores." They had candy and tobacco, toothpaste and that sort of thing. I was terribly fond at that time (never have really been since) of fruit-flavored Lifesavers. The Navy had them in cardboard cartons which had probably about twenty little boxes with rolls of these fruit drops. It was so nice in the middle of the night on the bridge. You can't smoke (because of the light), so you need something to suck on, allay the tensions.

We were in Subic Bay, at anchor, just sitting there waiting for orders (as usual, sitting around). I saw this tender a couple of miles away and I knew they would have candy. So I jumped in the wherry. I rowed, didn't trust the outboard. Sure enough they had these fruit drops. I got two cardboard cartons. Signed them out to the ship, one for me and one for any of the guys who wanted some. Rowed back, must have been over an hour. That was the measure of a longing. That was a long row across open water of Subic Bay, in the middle of a damn war just for some fruit drops. But it was a comfort. So you see the Navy even planned a supply of candy when and where we needed candy!

We went back and forth in the Philippines, but Tacloban, Leyte Gulf was the hub of our trips back and forth. There came a time when the Japanese threat was greatly diminished. By now Tacloban could be called a rear area like the other places that we had been. And we were even spared night time harassment.

But one afternoon the gangway watch came to me. He was sort of running things at anchor when nobody else was. He said, "You know there's a boat drifting by out here that looks like our wherry. I checked on our wherry and it's gone." First anyone knew it had even been missing, apparently stolen in the night. At anchor we kept the boat secured by its painter to a cleat on the stern—handy. Now there it was with its dumb little outboard perched on its butt end. And it was riding very high, no one in it. It was going out with the tide and wind at a pretty good clip. I was insane that somebody would steal it but I was more insane to think I could go swim out and catch it.

I just jumped out of my shoes, and into the water, and struck out for it. I thought, "Now that's going with the wind and the tide so I'll take about a forty five degree vector on the hypotenuse and triumphantly return with the engine roaring." So I did it. I was forty-five minutes!. Thought I'd never catch the bastard. It just danced on the water. So sprightly. It seemed almost always within reach. I got tired; I got very tired. I'd stop and tread water for a minute and get my bearings and then strike out again. Finally I climbed in and lay there on the bottom panting.

Do you think that thing would start? NO! We kept drifting out, probably not quite so fast with my weight in the boat. I would start it and get halfway, then it would conk out, we'd start drifting to sea again. Pretty soon there was a gang of guys on the fantail of the 353, hollering and urging me on. And the motor machinist's mates were hollering instructions. I'd get it started again and wipe it out. I got it started again and finally got to within about a hundred feet. It conked out. At that point the 1st Class Motor Mac took off his shoes and jumped in; he climbed aboard. A genius with outboards; he got that baby sneaked in alongside the ship. The enormous cheers!

What a dumb stunt! But to get our own, our very little own boat back. By now we would fight for the 353. This was just another instance of that lemon that we'd gotten in Manus when I was so generous to the whole little fleet of six sweeps.

As cautious as I was in ship handling and as a person, I had strange streaks of imprudence. This was one. It would have taken a long time, but the ship could have taken in anchor and made the trip. I was very foolish and could easily not have made it.

Mail was absolutely the best restorative for us all. It was stupendous how the Navy kept up with the mail. We did not have our mail at Manus after we'd left Hollandia. We got mail in Hollandia but by the time we got to Manus the Fleet Post Office was not expecting us. Fleet Post Office had us listed as Leyte Gulf, which was where we were headed; we went weeks without mail. I tried to explain this to letter-starved men.

First thing I did when we were anchored in the fleet anchorage at Leyte was to get a lift on a boat going into shore. Get the mail. We were greeted in Tacloban by again, this ankle-deep red, slick, cold cream-like, volcanic mud. Hmmm. It was a terrific slog, but we found the Post Office. Quonset hut after Quonset hut, open at both ends and stacked to the outer limits of these arched buildings with canvas sacks and outside more sacks in the rain. Honest to God, they located two canvas bags saying, "YMS 353." I will never forget that thrill that they could do that, and quickly. The bags were covered with green mildew. They were hairy and they smelled something awful.

The yeoman (ship's clerk) and I dragged them back down to the pier about a mile away, leaving deep tracks in the red mud. We looked around for a way of getting a ride on some landing barge or something going out into our anchorage. A PT boat was sitting there and the guy hollered, "Do you need a lift?" "Sure!" We threw the bags onto the focsle of that PT boat; they were just ready to shove off. We took off for our ship at forty-five knots. We were out to the anchorage in minutes. I've never been so fast on a boat in my life. What a thrill! Those things were fantastic! It's just a shame that, as Morrison said, there's no known record of their ever having sunk anything big, in anger. They were such beautiful boats, such beautiful, lethal, slick looking boats, such derring-do and dash about them.

They were actually very useful, among other things for doing what they were doing then, they were delivering messages around the bay, acting as ferry boats. New officers arrived to replace someone and out they'd go on a PT boat. At night they'd wander around looking for anything that looked Japanese and they'd shoot up landing barges of troops or barges with fuel or ammunition. They were useful in that way. They were sure a godsend that day we got our mail. And this was just the sort of unplanned way that everyone was so helpful all over. How can I say it stronger? Initiative and mutual concern saved us in the chaos of the forward areas.

That was typical of our mail, always an adventure. We took it, not into the wardroom, not into the galley, but out onto the fantail, as far aft as we could get in the clean and open. We ripped open those two sacks with knives, spilled the contents out and gave the deep six to those awful bags. And there were cardboard cartons covered with mildew and other colorful fungi. They just smelled of . . . dear mothers would send fruitcakes or cookies and two months later, after sitting for a month on a tropical island in the rain, they were delivered. Boxes of food went over the side awfully fast. They were followed with the jeers and boos from everybody, everybody but the recipient. I fortunately never got such.

There was official mail. It would come wrapped separately in a brown bag, a big rubber band around. We'd haul all those in, clean them up and take them into the wardroom. They wouldn't smell so bad by that time.

The yeoman kneeled at this pile of smelly paper and called out names to a hovering silent crew (this is the classical scene in war movies). We had our personal letters. Everybody found a private place after the mail came. And there'd be someone sitting under the gun up on the focsle. And there'd be someone leaning on the anchor. We would go to our bunks; the bunks were all full of guys, some with moving lips. In the galley, one guy at one end of a long table and another guy at the other end of the same long table. Down in the engine room, up in the yeoman's shack, up in the chart house. Reading the mail from home brought a silly look to our faces, sometimes a serious look, but usually a sort of sappy look. And I'm sure I had one of the silliest. I'd get those letters and could just make out, over all the other stink, that Chanel #5. I knew where it was from. What a moment! And it would last for many re-readings.

I was always amazed when we had mail, but one of the incredible things was how the Navy really did provide for us in one way or another even though we small craft were not on most people's lists.

The planning was excruciating. It was quite astonishing. One of the things that was most astonishing was the concept of a "tender." What a nice word. It's a class of ship, a tender. A repair tender could tackle anything. They had huge cranes. They'd lift an entire engine or gun on a big crane and haul it aboard and redo it for us while we anchored nearby or tied up alongside. They could do anything. Here's where the yeoman went to get a box of forms or the newest advisories on administration or manuals of instruction, new doctrine. All of this stuff came out and the tenders had them in multi-tuplicate for anybody who needed them. Same for medical staff, everything. This was just tending other ships. That was their role. They were extraordinarily busy places, day and night. Lights were always on. They had guns, lights would go out, they'd shoot at planes. When the planes were gone the lights would go on, back they'd go to work, aboard the other ships, aboard their own ship, down below.

They had medical and dental facilities. I once needed these on a sub-tender off Tacloban in Leyte Gulf. We'd been up north off Luzon and on one occasion, I had my only casualty of the combat days. I was watching a *kamikaze* plane coming at us. It came down on the port side and it just barely missed us. I was so scared that I didn't realize until afterwards that I was clenching my teeth so hard that I cracked a tooth, a canine. It began to hurt. I looked in the mirror; my God there was a huge crack. It was split in two halves, but they were solidly in place. Every time I touched it after that the pain was excruciating. I had to live with that damn thing for three weeks. I ate a lot of aspirin.

We got back to Leyte and I immediately went over to the sub- tender. We were anchored a very short distance away. Jumped in the wherry, rowed over there and said, "I gotta see a dentist."

"Okay." Up I went. A dentist and two chairs, they had a corpsman who assisted the dental officer in his work.

"What seems to be the trouble? Oh my, that's gotta come out."

I said, "I know that, there's no other way."

And he said, "Now look we've got all kinds of people coming in and this is not—there's no time for novocaine, just, just hold on here now, will you?"

"Yeah," I said, "Sure."

Ha, you know, tough guy. I'm not tough about that kind of stuff!. I don't like pain in the dental chair. But it was all right. I made it. He pulled out the pieces and rinsed out the hole. I was kinda flushed. He kept joking with me and I kept joking back.

And then he said, "Corpsman, bone chisel." Bone chisel?

You know, I didn't need an anesthetic after that. I just came that close to passing out. My head was swimming. No pain, no strain (our common expression of the day) and I barely was able to stay conscious.

He said, "Feeling a little woozy?"

I said, "Yup."

"That's okay, we're through."

He rinsed it and rinsed it. My face was still flushed. I began coming to again, with lots of pain. That was a hell of a hole. He packed it with cotton and said, "Hey come on back if anything goes wrong."

That was the tender. He cared for me. I don't know how you figure that out when a nation is at war. They have a dentist where a dentist is needed. It was an incredible feat to have all that stuff. It was Naval organization, but more importantly it was people doing more than their duty.

In Ulithi one time we got an AllNav (All Navy); came in the mail. The yeoman came running down and said, "You're on the list, you're on the list for lieutenant, you made full lieutenant." He had all the records, personnel records. Sure enough, I was going to be promoted to railroad tracks, two bars, full Lieutenant.

But I had to have a physical so I went over to a tender, there was a repair tender nearby. My ride dropped me off, I went up and I chatted with the pharmacist's mate, the corpsman on duty and he gave me the brief physical exam. Then the doctor came in and checked my ears and nose, heart and lungs and asked me if I had any trouble. And I said, "No." And he said, "You look in good shape." And I said, "I am."

Then the corpsman gave me a hearing test (whisper at fifteen feet), I passed that. Then he gave me an eye test; I passed that. It was so funny because for a year in Boston, when I was a Corpsman, Pharmacist's Mate Second Class, I had given thousands of these exams. I read off the first line with my right eye and then backwards with my left eye, "A-E-L-T-Y-P-H-E-A-L-T." Can't get rid of those damn letters. Then I said, "Hey look, no eyes." I covered both eyes, "the third line, O-H-C-D-L-F-N-T-C-O-C." Chuckle. He stared at me. I said, "I know that one. It spells, 'Oh see the elephant's cock.'" I explained to him how I knew and said, "I really can see them. Honest to God, I really can." He called the doctor. I told him the story. He was very serious; he didn't like joking about this sort of thing.

I said, "I can really see. Look I'll read out of your finest print. The Corpsman's manual has footnotes every other page." I read them. "Okay," he said. I never thought I would be joking about something like that on a tropical island two years after I'd been doing it day after day.

I got to Tacloban one time; I needed a haircut, but hadn't thought about it (aboard ship one guy did hair cuts—scissors and a hand-squeezed clipper). I saw a red- striped board on the outside of a shop in this bombed out, terribly desolate small city, deep in that red mud churned by jeeps and trucks and thousands of feet. Here were these Nipa huts. They were of coconut frond roofs and thatched sides, and there was this board painted red and white like a barber shop.

A woman was sitting on the porch and she said, "Haircut, haircut, meester?" There were three stools in there and some pans of water. They'd soap your hair, comb it, and cut it. It was very nice. There's absolutely no tradition, at least there wasn't when I was a boy, for having women cut men's hair. The barber shop was a sacred place. That was a capital M, "Manly" place. But here were the three women barbers, very cute. And all bubbly; smiles. They'd put their arm around you and hold your head. It felt kind of nice! Hadn't had soft female arms around me in months. These weren't USO show girls, they were barbers. It was a nice haircut. And I told people back on the ship, and they said, "Boy, you must be hard up, paying fifty cents for a gal to hold your head!"

I first came to appreciate government costs and this low bid stuff during the war, when I had to go through the receipts for materials received. We had what we called "ducks." They were markers for our sweep cable. It was two boards you could hold in one hand. One board bolted onto another with lag bolts. Like a big airplane a five year old would make, with two boards. In the front was a broomstick with a flag on it. An eye bolt in the nose had a piece of line which we secured onto our magnetic minesweeping cable. This bounced on the waves, and from the little red flag you knew where the end of the cable was; as much the splash as anything else. And they of course were expendable Two wooden boards pounding on the open ocean? They cost $80.00 apiece! We talked about it with our supply officer ashore and said there must be some mistake. "No, they're $80.00 apiece." And they just shrugged their shoulders, "Everything's that way."

We had all kinds of scenarios about it. This little guy, who, too old to be drafted, had been a carpenter all his life and he turned in a bid on one of these things. We had a vision of his garage set-up, with one saw, that's all it would take, and a drill. And he could just work all day long and turn out dozens and dozens and dozens. At $80.00 a smacker! After he got the low bid, he probably made all the little wooden ducks for the entire fleet.

This just got my dander up, because I knew I could make one in twenty minutes. They weren't even painted! I talked to the gunner about it, and he was furious. The gunner was our artisan, our craftsman; had a little shop with some tools. He just cussed out these "God-damned civilian war workers." This was a challenge to him. And he asked, "Where do I get the wood? Can I requisition some wood?" I said, "Hell, no! The pier is full of packing crates." We got a bunch of packing crates, we took them apart, he and I, with hammers and crowbars. We had some eye bolts and plenty of broomsticks; we were constantly wearing out swab handles. We had flag material. So, we did indeed make them, two of them, not one, in about twenty minutes. That was the last time we ever paid $80.00 for a duck. Damn swindle!

We did get awfully good at scrounging, the final resort to solve shortages. Every man on the ship just took it as a personal challenge the minute he went ashore to come back with something that nobody expected. We particularly looked for large ships that were low in the water, namely, loaded. Small ships gathered around them as they anchored. "Got any vegetables? Flour? Lube oil? Gaskets?" They would say, "Come on, come on." It was a way of life. They responded well. At a time like that, when we were in a forward area, nobody worried about requisitions or signatures, everybody pulling together, quite an astonishing thing. It sounds so corny. There was a job to be done. This was war; we needed the stuff; no one planned it but we got it.

We even had a chance to reciprocate. One time we escorted a convoy from Ulithi to Kossol Roads, just north of Palau. We were herding a flotilla of LST's that was going to anchor there in the lee of the Palau archipelago.

And they were going to go to Borneo I think. There was a patrol boat down there, a PC, which had been there for two months, just stuck there, patrolling. And this was not, this was not, I repeat, not a place where there were supply ships where you could get stuff. It was really bad. There were only combat ships heading right through.

The PC came alongside of us. "Have you got any tobacco? Have you got any flour?" And it sounded so familiar but so pathetic coming from our peers, you know. They knew we'd come from Ulithi were there where all those tenders and concrete supply barges. We took their mail; "Yes, we'll take your mail." We gave them our latest code books, promotion lists and one thing and another, some medical supplies. These guys were in trouble. They were really in trouble. They were out of everything. We gave them all the food we could; we were only a couple or three days back to Ulithi and a chance to get more supplies.

In the Philippine Islands, civilians suffered the Japanese occupation and then the fighting. At night families would get in their canoes, no one would see them. They could quietly paddle anywhere. They would come alongside and whisper, "Hello, hello." These weren't organized beggars, they were just people. This woman sitting down in the bilges plaintively called up, "Soap?" Oh, break our hearts! My mother had given me a marvelous little sewing kit, with needles and all kinds of thread and buttons. One woman accepted that and cried, thanking me for it. She had nothing like this at all. I gave her all of my skivvy shirts that I could (T-shirts, white T-shirts). These would fit her husband, be loose on her and gowns for the kids. I could get more, I could get more. We could draw those as Naval small stores; the officers paid for their own. But that was just terrible! Terrible that we had to give them some food and clothes. This was real pathos. War.

That trip from Guam to New Guinea took us farther and farther into forward areas, although we were going away from the Philippines. It was also a rude lesson in self-sufficiency for the crew and the skipper. The trip gave us a jolt. I don't remember just what day it was, after we left Guam. We were approaching Ulithi, where we were going to stop, fuel, and provision. This was routine escort duty, boring.

All of a sudden, the engines stopped, dead. All of them. There was a ghostly silence and the waves slapped and each voice on the ship could be heard, and there were no lights, no radio no coffee maker, no radar, no sonar. There was no motor going, "Click-click" every time the wheel moved. I'd heard the first little faltering, the first hesitancy of the engines. I knew every sound, every engine; we had several. I was in my cabin, when a voice called down the tube, "Skipper."

"I'll be right up." We did kick in one engine; we limped. Called the commodore ("Sara"), "Engines out, attempting to repair. What are our orders? Over."

"Merrimac 353 from Sara. Rendezvous with convoy in Ulithi. Out."

We watched the convoy sail south over the horizon. We were alone—the sea and the sky. Time crept by. I was in the engine room where the black gang worked like the possessed. Very serious bunch, there was no humor here. We had taken sea water in the fuel we'd gotten from our friendly LST in Guam. Diesel injectors had to be cleared, blown out, one by one. All the fuel lines had to be drained, cleaned out, cross-connected to our reserve tanks. As I recall it took several hours. And then the other engine started. What a sweet sound! There were cheers all over. We steamed slowly. We got them all started. Recall that this was the first area where we had had a red alert. An air alert. We hadn't seen the planes, but we were very, very nervous. This was not so far from where the "Great Turkey Shoot," off the Marianas had taken place. But that was a month or two before.

We picked up the ships at Ulithi by radar. Then in the morning we nosed around and found the channel, Mugai Channel. And we reported to "Sara." This adventure helped us . We matured, we were bonded. We had survived alone on that empty sea. My God! Also we were rid of "Sara" for all those hours and hours! Getting contaminated fuel was not a shining hour for our supply system.

It seems like a good place to digress for just a minute. It would seem, in this chapter on supplies, that I am proclaiming the moral posture of stealing and lying and that I call shame on the Navy that could not care for us the way they did battleships. I want to briefly address these.

We lied and we stole and I bent the authority, but, (1) under trying conditions, truly trying conditions, and (2), for the good of the damn Navy! This was never done for personal good, anybody on the ship. I know of no stealing, lying, cheating on the ship at the personal level. I never did this sort of thing when we had duty off the East Coast where we were solidly supported with the entire U. S. of A. supply system. Great.

We understood that the Navy could not keep track of us and send us a tanker every time we needed fuel, the way they did an aircraft carrier. And we were impressed by the sharing of supplies and noblesse oblige of men on the larger ships. Actually we were a fraternity, and it was for the good of the Navy.

I was truly concerned about war profiteering, recalling the $80.00 wooden floats, and I saw other prices that were inflated. How much would a three inch shell cost? A single twenty millimeter shell? We sprayed the air with these, staggering amounts! The ship itself was worth a million dollars. Sailors in general had bad feelings about medical deferments of civilians, and people making a lot of money stateside. But on the other hand they all knew people in shipyards and heavy industry, driving buses, a dad or a sister. We knew we needed shells and we knew we had to have them regardless of cost.

There was the storied island of New Guinea ahead of me. I saw it on radar first; the spine of the mountain range showed up 60 miles away. And then we could see the line of green peaks in the morning early. We could smell land, the rotting vegetation, and that rank, raw-earth smell of the tropics. There was the usual coral of the fringing reef, but then mangrove forests, deep, dark green clumped vegetation, crowding the shore. Inland the tropical rain forest. We could see the opening into Humboldt Bay, a wide opening. "Sara" and all the ships slowly filed in and we could see the mountains on either side of the opening into the bay and then we could see Hollandia Harbor. It has another name now; that's the old Dutch name of course. And, there was a signal tower, and there was much blinking, a lot of radio traffic. Now we'd finished this incredible journey, chasing the combat zone all the way from Charleston, in a more or less continuous swing onto New Guinea.

How we hated "Sara;" never met the man, the convoy commodore. But one of the things that will stick in everybody's memory on that awfully long trip from across the Pacific down to New Guinea, was "Sara." I've referred to it obliquely; everybody, everybody on the ship, had a strong distaste for "Sara," or whoever it was on the bridge of the flagship that sent out all the orders. It was not just the tone, "THIS is Sara;" it was more than that. It was "So and so ship, PLEEEZE keep station more attentively. I repeat." Or, "Four of the vessels have not submitted a noon position report. I repeat." "Attention all ships in the convoy. At such and such an hour all vessels will have General Quarters drill. I repeat." He was fussy. He was fussy about everything. And there was this awful snooty tone in his voice. And everybody on the ship went around mimicking him and making fun of him, but also being kind of bitter about him. We were all doing a good job. And, as far as I could see the station keeping was quite good; no one was badly out of position. We guarded him. But apparently this man wanted everyone to toe his line. And he talked down to the commanding officers who served under him, unnecessarily.

We did finally pull into Humboldt Bay, New Guinea. Exciting. Then as we all steamed into this broad bay, we slowed in accordance with flag hoists on the flagship.

Suddenly on the radio, on TBS, "All ships this task group, THIS is Sara. LST's detached, report ComPhib. Escorts, you will receive orders. You are hereby detached. Out" Silence. We waited for more. Silence.

Then suddenly another voice, a loud, very clear voice: "Fuck you Sara." Then from every ship in the convoy, every radio man, every officer manning the TBS, joined in the chorus not only "Fuck you Sara," but every known possible derivative of the idea, every cuss word anybody knew, every obscenity. Laughter, riotous laughter turning to derision. The commodore spluttered, "No unauthorized traffic on these frequencies." More, "Fuck you Saras" from everybody, more glee and derision. Then silence from the commodore. What a scene it must have been on the bridge of the flagship. What a scene. What a silence. There was sweet, sweet revenge; what a fitting final episode for this very long trip and what a genuine coming of age the entire crew had had in that one long voyage.

We now knew the arrival drill. We went to a pier right away and we found things out. We talked. We listened. We made friends. I went to the Port Director and got a local chart which we didn't have. We found where the fuel barges were, the water barges, where we would get mail. Engineers fanned out and got some spare parts. The cook found the food warehouses. At night the officers went to the Officers' Club, the noncoms went to the Noncom Club. We went visiting other ships. We got to know the local drill in a hurry and the supplies rolled in. We got mail, not much, but some mail. No orders from headquarters. We were surprised, but that's alright, things take time. We slept, we slept very hard. Our corpsman got atabrine for us to start that regimen. We were in deep malaria country now; the air was just thick with mosquitoes.

We waited—no orders. The radio was manned all day. Signal tower watched.

Didn't the Navy know where we were, where we were going next? Day after day—no orders. We sat.

I took off one wonderful day there in New Guinea. I just took off; told people where I was going. I was heading east to the coast, which was some few miles, and on a trail out to small villages. The trail was a crude path on the forest floor that faded out. The notion that I could be attacked by wild critters was not there. I felt at home. I did have a Marine knife, long, strong, and the handle was a three-holed "brass knuckle." No training in its use!

What an incredible biologist's dream this was! This was absolutely undisturbed forest at such a short distance from the very disturbed area of Hollandia. The vegetation erupted everywhere. There was a green layer on the floor; no dirt showing anywhere, just this soft carpet. There were tree trunks towering above a corrugated forest floor; the decaying old logs were green covered, long lumps. Then way up in the sky this canopy of leaves came together so it was dark and it was quiet and it was misty.

The forest was so dim. How could plants grow in this light? But in places bold shafts of sun spotlighted the floor—a giant tree had fallen. Here was massive plant growth. And, through the dim leaf canopy there came showers of light—sparkling drops of light that seemed to fall silently onto the soft floor.

The bright butterflies, the classical tropical rain forest butterflies; there they were. The birds of paradise floating in the canopy. You could just see their bright, incredible feathers. And there was a bird that kept hollering, "Keokuk, keokuk." What in the world is that? I followed that call. Finally I found on a branch, a little brown bird, about the size of a robin, just bellowing, "Keokuk." And I thought to myself, "Ah hah! No bird of paradise this, but what a voice to make up for it. One way or another they'll be found by a mate."

While my being was absorbing the majesty and numbing beauty of the forest, the biologist's mind forced me to see the structure and function. Growth and decay in the tropics are apparently in balance, but rapid. Garbage disappears overnight! Since then I have come to recognize massive plant systems as giant water pumps; water up through the roots and out from the leaves into the air. And that hot, moist air forms instant clouds, major water sources around the world. Lately I have been impressed by satellite photos of the South Atlantic. As the Earth turns, huge bands of clouds drift from equatorial Africa across the Atlantic to the Gulf of Mexico; here is the potential water for the great corn belts of the United States.

I finally got to the sea in the mangrove jungle. Marines tried to land on the mangrove jungle. They couldn't make it through much mangrove forest, not even with mechanized equipment. The branches send down roots into the water. The roots grow into trees. The trees have branches. They send down roots from the branches. It becomes impenetrable. By wading up to my ankles, my knees, my waist, I slogged through these, fought my way through and got to the sea. What a place!

There on the mangrove roots, out of water, perched some fish and I said, *"Periopthalmus"*. I knew the name of that fish. I'd seen pictures of it in the books. They have eyes that pop out, huge, big pop eyes. They can see in the air and in the water. When I approached them, they just let go, they did not dive, they just let go and fell in the water, swam away. What an incredible thing. There were crabs, oysters, snails on all these branches and roots. I didn't know their names then.

Being in the forest interior was almost a religious experience. This was a cathedral-like setting; it's the only way to describe it. You expected the rich sound of organ music but instead it was just the utter stillness with the occasional bird call and the fluttering of wings.

We made lots of new friends in Hollandia. The Officers' Club was a marvelous place. It overlooked the harbor and the forest and the mountains. Everywhere there were mountains. There were two three-thousand foot mountains around. The Cyclops mountains rose up in the middle of the island. It was just stunningly beautiful. But it was a dank, dank, hot sweaty harbor. We could spend a whole evening at the club; met new people and drinks were only a quarter, that was standard all over.

I got to know a couple of the skippers who had been with me on the convoy and we wondered if Sara had ever told anyone about our being here. Why no new orders? We were all resupplied and ready in one day. You see, Naval administration was more than food and fuel; it was another matter to keep records of where each ship was and the men on them.

We were directly astern of this huge ocean-going tug alongside the pier in Hollandia. I was invited to lunch in their wardroom one day. It was so spacious to a YMS guy. The conversation was sparkling; and the duties of tugs so different from ours. But it was remarkable that they were not at sea, had in fact been here a couple of weeks, rubbing the pier.

One of the young officers told me the story. They were confined to port, no leaves for the crew, until a Board of Inquiry could complete an investigation into alleged misconduct of the Commanding Officer.

Their skipper was a "mustang," commissioned directly from the ranks; he had been a Chief Bosun's Mate on tugs, Regular Navy. Such a man knew tugs, their mission, in all weather; he knew small boats, deck gear, winches, rigging, cargo handling. A natural for command of such a tug. He was also described as a boor, unable to deal with the young reserve officers; and he was a drunk, ashore and afloat. This was the wardroom view that I heard. He was indeed an "officer but not a gentleman."

I asked, "How did the investigation happen?"

"I'll tell you exactly how it happened," one of the officers said. "I knew in less than a half minute into the preliminary hearing, right in this wardroom."

The very senior Captain of a very senior group of officers opened the hearing with his charge: to investigate allegations of misconduct and failing to perform assigned duties afloat, etc., Orders came from the Department of the Navy, Secretary of the Navy; the allegations were from the senior Senator from the State of Carolina.

My host chuckled and told me that he immediately raised his hand for the attention of the Board.

"Sir, I am sure I can tell you how this came about. I accept the blame. The Senator is a close friend of my mother. I wrote her a letter critical of our Captain and I stamped my letter and initialed it censored. This is surely the reason for the convening of this Board. I apologize for the inconvenience."

Inconvenience!

Each man and officer was questioned. Records of each reviewed. I never heard how it was resolved. The officers supported the charges and each other. They were bitter toward the Captain. The crew was apparently reluctant to testify against their Skipper.

I had some hunches about the affair.

The Captain could indeed have been a boor and a drunk; he surely lacked in leadership. He was also probably as good a man as they could find for tug duty. A very demanding service.

The young reserve officers were probably as able as most; they were certainly young gentlemen of fine families and the best schools. They of course had but a very, very small fraction of the Captain's experience at sea and on tugs.

The irresponsible letter by one of these officers was not the act of an "officer and a gentleman." The stamp and initials on censored letters was the word and honor of an officer, an anachronistic but surely a noble tradition. We were secure in our "word of honor." It was real and not taken lightly. It worked. Not this time, however.

Did this officer damage the Navy by not acting with honor?

Would more damage have been done, perhaps mortal damage, if this man had not blown the whistle to his mother who was so highly placed?

It would come as no surprise to me if the Navy had finally reassigned both the Skipper and the young officer to shore duties, with a censure in their files. Tragic to the old Skipper for whom the Navy was life. Trivial to the young officer who may indeed have gone on to high places. He was that kind of guy.

There is no way you could know of the caste system that held in the Navy in those days. Regular Navy vs. Naval Reserve. Among officers, the Annapolis graduates were Regular Navy, "trade school" boys. They were making the Service a career. You won't believe this but I never met one! All the minesweeping officers were reservists, not intending to make the Navy a career. We were the "Feather Merchants," far beneath the noses of the regulars. Not only sweep officers, everyone else I ran into were reservists: the amphibious fleet, PT boats, sub-chasers; medical, dental, supply, chaplain types. All reservists. This was the bulk of the Navy.

Quite properly, the Annapolis officers were on the larger combat ships and in shore command units. They were highly and widely trained, and needed visible berths to forward their careers. They were the backbone of the service; we were the rest of that body. The reservists were usually trained in one area and were turned out in a short time. The Navy consumed ensigns! For instance I went to officer's school for two months at Throg's Neck, the Bronx. Then another two months in South Boston for small craft school. I was limited to small craft, such as a sweep or sub- chaser. Regulars came from one college; we came from hundreds. For the Navy, a college degree was the major qualification. Of course, one's record, character, health, etc. were considered, or so they said. The non-college officer like the tug Skipper was the rarity. Neither group could claim more brains, bravery, or dedication to duty; the regulars surely had more dedication to the Service.

We all bled red.

We still hadn't gotten orders and we were almost a week in Hollandia. I went to see a friend, a resident supply officer. Why was I concerned? You know, the kid that was never late for school.

"Where are the headquarters?" I asked. "We never got orders; I have to find operations."

"Up in the mountains. Operations Office is up there."

"Where is that?"

"About twenty-five miles inland up on Lake Sentani." A fabled, tropical mountain lake.

He should know. He knew everyone and he knew what they could do for you. He knew what he could do for them. This was a classical supply officer. He had been there for several months; he was probably destined to be there for the entire rest of the war. And so he built himself a palace. He commandeered two or three tents and a Quonset hut and he had these all furnished as a bunk house and as a bar. He had two or three deep freezes, refrigerators. And a storeroom full of all the possible goodies he could want and so he had all the women he could want. He confided that going home would be a nightmare. I saw photos of a wife and daughter who were "coming out" at the country club next summer. Here he was a stud in a stable of fillies.

Right across the street there was a compound surrounded by barbed wire and in the corners guard towers up on posts. These were the WAC quarters (Womens Army Corps) and they were almost regarded as prisoners. They could not leave except in a truck or in a vehicle of some kind and that vehicle must have an armed guard. This was the fate of women in a relatively forward area. They worked in the offices and warehouses. They did hard work in that terrible place and with that kind of restriction apparently necessary. I can see why. This was a pretty horny bunch of guys, thousands.

He put me onto finding Lake Sentani headquarters and I did that the very next morning early. I had no trouble getting a ride, thumbed a truck with a couple of other guys and a bunch of crates; I piled in with them. It took hours to go the twenty five miles.

There was deep mud. It was slick like grease and the road was rutted and ragged. It had been blasted and bulldozed; there were rocks and logs, and these caused endless delays; there was an incredible amount of cussing. But here, as we saw so many times in the forward areas, I saw problems that in civilian status would have brought out the worst in men and here it brought out the best in them. They cussed not each other, they cussed the road, they cussed the circumstances. But they helped each other. They teased. They were sarcastic toward each other. But there was no rancor towards other people and when a truck stopped, everybody including the trucks behind, all pushed. Up we went, this endless centipede of cars travelling up to the headquarters. We got there finally. I jumped out, thanked them profusely, I don't know for what.

There before me was Lake Sentani. A blue lake. The sky was flecked with fleecy clouds on blue. I felt everything about me was blue. The mountains in the distance were hazy blue. And below, down from where I'd come, it was blue gray toward the sea, which was a blue in the very great distance. On the coast behind me was that miasma of dank air; here was clear sunny air. The breeze dried me and there wasn't a mosquito.

I looked uphill and saw this troop of buildings going up from the shore of the lake and at the hilltop a home of some kind. That was Douglas MacArthur's home with his wife and child, this young son and his nanny. And everywhere, headquarters of everything. Every building had a name (Com this and Com that). And the triumph of the Quonset hut was no more in evidence anywhere I had travelled. Quickly erected, formidable buildings, offices and warehouses. This was a place where the brass did their planning; the planning for the Philippines' invasions was done here, for instance. But so peaceful. Above the muck. Serene.

I finally found someone in charge of the Naval traffic in Hollandia Harbor and I reported in to him and it was news; it was news. He was stunned. Sara had goofed.

"Well, where are the LST's?"

"Oh, they're down there."

"Oh my God," he said, going through some papers, "you're supposed to be in Leyte."

"Well, we're not."

The wonder of it was they kept track of ships as well as they did. I waited and I waited. There were conferences and there were more conferences. Finally, "Get ready" he said, "get ready to move out. Too late for Leyte, probably go to Manus in the Admiralty Islands. You'll hear from us."

I got back down and in a day, it was almost with relief that we got orders. The six YMS's that were Sara's escort made the very short trip to Manus Island which is just north of New Guinea. We made the trip and we were suddenly in a very different environment. This was a vast staging area for the ships gathering to go on to the Philippines.

We were still in a very rear area. At Guam, it had been a month or two since it was secure. At Ulithi it was only a month. Been a half a year since the invasion of Manus so this was a well established place; this was a complete base and a harbor that had hundreds of ships. Seadler Harbor. We went to Lombrum point, which was an area of small craft; it was here that we staged for the Philippine invasions.

Neptunus Rex:
Crossing the Equator

Domain of Neptunus Rex

KNOW YE, that Lt. (jg) R. V. BOVBJERG on the 15th day of November 1944, aboard the U.S.S. Y. M. S. 353, Latitude 00 00, Longitude 140 E, appeared into Our Royal Domain, and having been inspected and found worthy by My Royal Staff, was initiated into the Solemn Mysteries of the Ancient Order of the Deep. I command my subjects to honor and respect the bearer of this certificate as One Of Our Trusty Shellbacks.

Davy Jones
His Royal Scribe

Neptunus Rex
Ruler of the Raging Main

8. Neptunus Rex: Crossing the Equator

Sophisticated friends are dismayed that I carry the card of a Shellback of the Domain of Neptunus Rex. I crossed the equator at sea and was initiated in a major and traditional event. The YMS first crossed the line north of New Guinea, south of Ulithi, 140 degrees longitude. At that moment, which we got from the chart and our working navigation, we had the ceremony. There were five shellbacks, old timers who had crossed the line, and thirty polliwogs, myself included, who had not. This was initiation time. The shellbacks were ready, and they had our approval; they certainly had mine, and they talked with me about their plans.

There was a drawing of numbers for the order of induction. It was important because they needed shellbacks to run the ceremony, and there were only five of them. Since each man inducted into the domain became a participant with a chance to paddle the captain, a very low number was the best. But they came to me, the captain, for the first draw. They had a hat full, a baseball hat of little folded pieces of paper. I suspect there were thirty of them in there. And, lo and behold, I drew number 30. Surprise! I never did see the other twenty-nine. I didn't want to. But shortly after that I was, by appointment, accosted on the bridge by a piratical character who had come aboard over the bow, from Davey Jones Locker in the depths. I don't know how he did that. We were under way escorting a convoy, were manning a position on the screen. And he came in a costume; ragged dungarees, an old shirt, a big sword, a bandanna around his head and a dagger in his rope belt. A black moustache and a little beard were painted on his face.

I met him in my full uniform and he demanded that I surrender the ship to him, which I did. He accepted the surrender and I took the conn on the bridge; we had people in the engine room of course; we were manning the ship. We were at war. But, one by one, they went through their induction ceremonies out on deck, on the fantail mostly. And needless to say I was the last one so I saw everybody get inducted. I knew the drill when my time came.

There was a royal court: Neptunus Rex and his Queen, a royal baby, usually the fattest guy on the ship, in diapers, and a bow in his hair. A royal barber was ready.

The polliwog was brought before these people, and in a kneeling position, bowed to the royal court. He was offered punch. A royal punch, King Neptune's royal punch. I found out later that it was made basically of salt water with a large slug of diesel fuel and then a shake or two or a spoonful or two of every condiment in the galley, from ketchup to salad oil to pepper, thyme and oregano, name it. This was kept mixed by continuous shaking. They kept it sort of frothed up. The polliwog was given a little cup and he drank. I did not realize early on why all the men went to the rail and vomited. I vowed, "No, sir. I would not do that. The Captain doesn't vomit."

Each was asked, "Oh, you like it?"

"Ugh," he would say.

The polliwog kept getting more punch till he claimed to like it.

The charges were read. This was a regal court, and the punishment always fit the crime. For each man they made charges that were fitting but a little bit cutting sometimes. Sometimes very cute, sometimes kind of silly. Mine were shooting stars, throwing ducks in the water (ducks are the minesweeping floats), not paying the crew, reading the crew's mail. More charges than the others.

Then sentence was read. To start, there was a fixed sentence for everybody. Everybody had to do three things. First you kissed the royal baby, who had his big belly hanging over his diapers; a belly smeared with diesel fuel or oils and then graphite. When you came kneeling to kiss the baby he took your face and shoved it into his navel;, you came away with these awful greasy, black streaks and spots.

Then the royal barber cut your hair. They'd constructed a huge canvas tank, made within a wooden frame, filled with sea water. There was a board at the top where you sat; your hair was shaved off raggedly. And this was where they really took it out on the guys who fancied slightly longer hair. With the final snip on the hair the polliwog was pushed back into the water hollering, "Shellback! Shellback!" He was dunked till he was coherent!

Finally everybody also was assigned to a crawl through the canvas, garbage or slop chute. This was about thirty feet long and sewed together very carefully; even a large man could wriggle through it using elbows and knees. Crawling, not in a hunched position, but in an absolutely prone position. This meant that you went through there very slowly. Meanwhile any shellbacks could whack at you all the way through. There was only one rule, you didn't hit people in the head. At the end of this the engineers had a fire hose rigged, so as you came out you were blasted back in. You had to fight your way out. But by that time you were cleaned of the garbage; I mean garbage. I mean potato peels for a week. Bacon grease. Coffee grounds, old cabbage. Whatever was gathered for a week in the tropics. It was frightful, just frightful. Nevertheless it was a final punishment. One came out hosed, clean, a shellback.

Then each man was assigned specific tasks, tasks to do as punishment. For instance there were two guys standing in the officer's head with gas masks on. And when an officer came in they were supposed to say, and they did, "Officers stink." Not exactly Hollywood level humor, but broad. There were some Irishmen who had to beat on upside down buckets with sticks and sing "Macnamara's Band" and other Irish songs as they marched around the deck. Some guys swept the deck (most of these were engineers, the deck gang wanted to get even with them for never having to do this), and they had bracelets and necklaces of heavy chain and heavy cables. Absolutely back-breaking things that went "clank" as they swept the deck. Two men stood on the coiled drum of sweep cable. They had beer can field glasses scanning the horizon. They called out lookout language, "Land ho!" "Thar she blows!" "Message from Sara!"

There was no escaping this punishment. One after the other everybody got through, but the real crux would come when the skipper appeared, number thirty; then everybody would be a shellback but the skipper. So that as many people as possible could take their turn whacking at the skipper.

"Other than that Mrs. Lincoln, how did you enjoy the show?" I recall that old line. I could not help but be tickled about this long afternoon; yet there was a certainty that my time would come. It came and I was sort of ready, having seen all the others go before me.

My bow was so low and I smiled so sweetly; the royal court brooked no smart talk. Then on my knees about to be offered the punch I remembered my vow, "The captain will not vomit." And I did not vomit. But I must have swallowed that stuff three times, between surges of blind nausea. Then in a choking voice I had the magic response ready. When they asked my opinion of the punch I responded, "Loved it! Can I have the recipe?" They replied, oh so deftly, "In that case have some more!" They had me. The captain did not vomit, victory. I never will know how I held that brew down. I have always had a "tender tummy," weakest part of my body. I certainly was not "well" for half a day!

I kissed the baby. The entire crew could not wait to bellow.

My hair was so short, it was decided that it could not be cut. But I did get the dunk in the pool of salt water. And I knew this drill. So I stayed down a long time and then suddenly reared out of the water hollering the "shellback" so clearly that I was not redunked.

My tour of the garbage chute was misery. I expected the beating by all hands but I did not expect the claustrophobia; it was almost impossible to make progress. And I know that I lost some of my royal punch, adding to the vile road. The fire hose was a blow at the end, but how welcome! The guys all grinned and banged me on the back. I staggered and laughed back. It was a moment.

I wonder why this is such an important ceremony. I have to this day a card, issued by me in the name of Neptunus Rex, an official card printed by the Navy, on entering the domain of Neptunus Rex, with the date and the longitude. My Naval files carry an entry on the event. One carries that card, one carries that card very carefully. When the ceremony was over I was given command of the ship again and Neptunus Rex and his court disappeared over the side in some way (on ropes?) and were never seen again.

There was a kind of comradeship in the crew that there hadn't been before. This was something done together and there was a "bloodletting," an official raucousness; this was rampaging hormones, and aggressiveness of a bunch of guys in their youth, on this tiny ship.

There was pride. How many folks get this opportunity? So it is with other ships from what I've heard and from what I've experienced. I've crossed the line several times on ships. Sometimes with no ceremony. Twice I've been on oceanographic expeditions and we did hold a ceremony on each. One was in the Indian Ocean and one was off the coast of Ecuador in the southeast Pacific. Here was this highly educated scientific crew, mixed with the ship's crew, getting together, drawing up all these specifications and plans and conniving in the same roughhouse approach on the part of both sides And lots of shouts and lots of laughter. Maybe it's good. I always took a dim view of fraternity hazing, especially those that got absolutely mean. But this affair was a good thing. Camaraderie; family.

Unending Terror: Combat in the Philippines

TRACK OF YMS 353 IN THE PHILIPPINES FROM LEYTE TO LINGAYEN GULF AND FROM LEYTE TO SUBIC BAY, VIA SAN ANTONIO

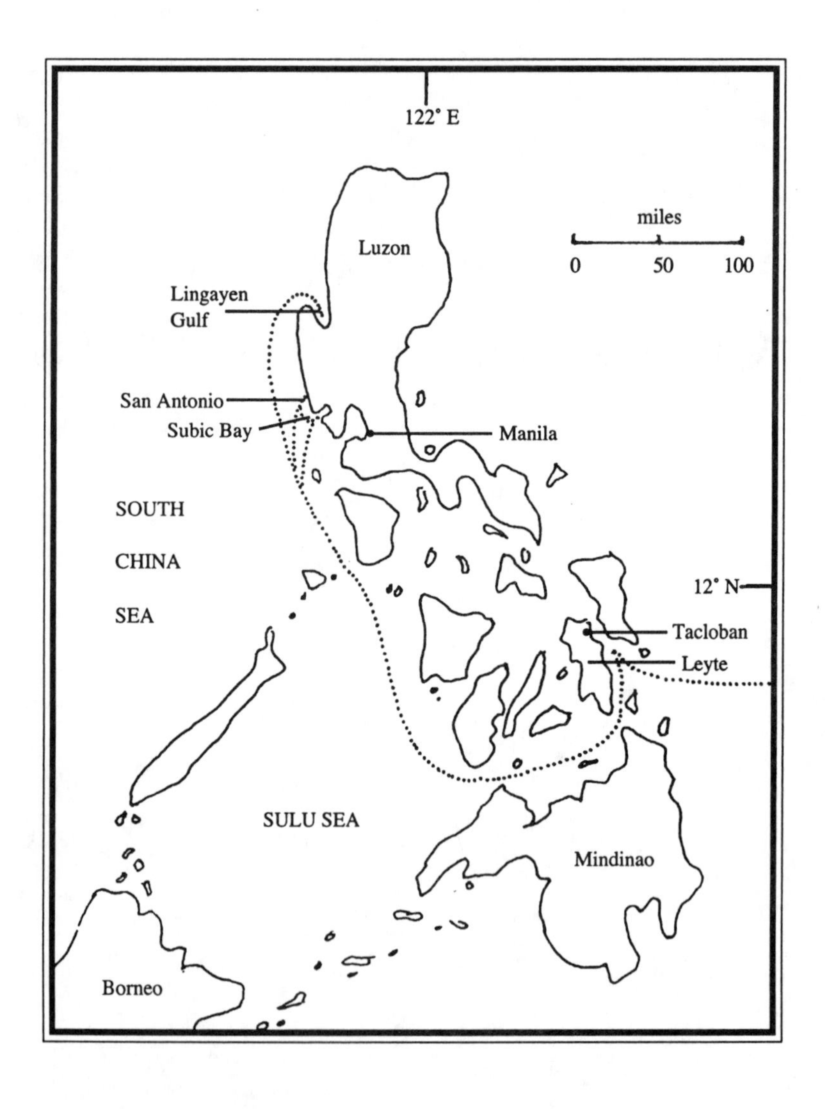

9. Unending Terror: Combat in the Philippines

We spent Christmas, 1944 quietly, and with no "Ho-ho-ho." We were on the way to the Philippines and that was the stage. The week it took us (the task force of minecraft) to go to Leyte Gulf from Manus Island was our plodding entrance from the wings onto that stage. We had been chasing the Pacific War for four months. On every island, a month or two behind. And despite Tokyo Rose's predictions, the passage from Manus to Leyte was uneventful.

We were about seventy ships in the minesweeping unit. There were forty YMS's; we had the AM's, which were the so-called large fleet, steel hull minesweeps; we had some DMS's, Destroyer Minesweepers, and APD's, Destroyer Transports with Underwater Demolition Teams. I had the invasion plans to Luzon essentially memorized. Our ship was not listed, but the unit, the minesweeping unit, was. And for every hour of every day the plans listed what courses we would be on, and where we were going, and what the contingency plans were. Resupply and refueling, everything.

In this invasion we were going to make the end run from Leyte on the east side clear through the Philippine Islands around to the west side, about a hundred miles north of Manila on Luzon Island. This would be Lingayen Gulf and would be the crucial battle for the Philippine liberation. (Check map)

We would resupply at Leyte Gulf be there for four days before starting the long trip through the completely hostile waters of the Philippine Islands. Arrival time at Lingayen Gulf was three days before the invasion, H-Hour of S-Day. We were to clear the mines and beaches. This was an enormous minesweeping job. Lingayen Gulf was hundreds of square miles. And all less than fifty fathoms deep. We were to get close cover by the Bombardment Group, and the air protection of the Lingayen Fire Support unit. The planes of the Pacific Third Fleet Carrier Task Force would provide Combat Air Patrol (CAP: great acronym, CAP). And they were to bomb airfields on Luzon and on Formosa (now Taiwan) which was only four hundred miles almost straight north.

Each day on the way to Leyte we held dawn and sunset General Quarters; all guns were manned. This was taken even more seriously than it had ever been. More and more serious with each day closer to Leyte Gulf. Leyte had been invaded two months before and there was still fighting there. This was the site of the famous sea battles: Surigao Straits, Samar, those which ended the large-scale naval threats of the enemy. But the dreaded unthinkable weapon, the *kamikaze,* was being introduced in Leyte Gulf at this time.

This huge gulf at Leyte was our landfall on 29 December 1944. We found our way into an anchorage of hundreds of ships. San Pedro Bay, our area, was an anchorage specially reserved for small craft such as we. I want you to read from the log of 31 December 1944, a day in Leyte Gulf. This will give you the feel of the place.

> **0000.** Anchored as before in small craft anchorage, San Pedro Bay off Tacloban, Leyte, Philippine Islands. General Quarters. Secured at 0700.
> **1028.** Underway to go alongside APD 10. Tenders to fuel her with 2,000 gallons of diesel fuel from our tanks.
> **1140.** Moved alongside fuel pier.
> **1157.** Commenced pumping fuel.
> **1410.** Secured fuel lines.
> **1424.** Underway, San Pedro Bay anchorage.
> **1600.** Anchored.
> **1700.** General Quarters for sunset.
> **2020.** General Quarters for an air alert. Flash Red.
> **2030.** Secured.
> **2315.** General Quarters on a Flash Red.
> **2340.** Secured.
> **0000.** Anchored as before in small craft anchorage, San Pedro Bay off Tacloban, Leyte, Philippine Islands.

That was a day in Leyte Gulf, typical of the four that we spent there. Twenty minutes after the last General Quarters and before the next one it was a "Happy New Year, Everyone." Not a lot of "Ho-ho-ho" here either. Our days in port were routine by now. Provisioning. Food, fuel, ammo, water, mail. Ship-to-ship, shore to back. We met officers from the other ships. We had meetings. We waited. We wondered.

Got to know the officers of a destroyer sweep; they were anchored not far from us. I went over there to have lunch. This was an old World War I four-piper which they'd cut down to two stacks. They were fast sweeps. They'd go better than twenty knots. We had lunch in their wardroom there and it seemed palatial. We had a feeling that it would be nice to have these "big guys" with us, with all those little YMS's. They had a pet monkey. It was shocking. It was ludicrous; it was charming. I don't know where they got it. It was crazy; it was so out of place.

Our ship was finally completely ready for the 2 January take off for Luzon. And Lingayen Gulf. And the war. Our guns now were never covered with their canvas hoods. We had been bombed and strafed at anchor in the Gulf. We were pestered through the night by "Washing Machine Charlie" who buzzed around half the night keeping us awake. Just before sunset, which was such a good time for *kamikaze* attacks from behind the mountains to the west, dozens of landing barges made smoke. These were large landing barges and they had smoke pots out on their fantails which poured out clouds of vile smelling, incompletely combusted diesel fuels. They weaved in and out and around until we were all covered over. We were all night in this choking, blinding stuff. But, covered, and invisible from the air. Bombs falling could only hit anybody by chance. Of course we were sure it would not be us. We . . . we hid. We were not allowed to fire. Good, a good strategy. We slept, but not well; the skipper actually lived on a bunch of half-hour naps.

The *kamikaze* inspired terror. This was to become the major weapon against us. It had an unreal aspect about it. We could not understand it; the American mind could not handle suicide. And they were very effective. The Japanese fleets did not beat us. Not all their shells and bombs or torpedoes. But a single plane, aiming itself, hit hard and ships burned. Many sank, and the casualties were heavy. There was carnage. Burned people. It made us sick and we were angry and confused.

I recall that night before we left, this was a sober moment. Were the officers and crew ready? Was the Skipper ready for this? It seemed like the war was with us in Leyte Gulf, but the answer was clearly, it was not. This was already becoming a rear area. But I did decide we were ready. For more than one reason. First: we were four month veterans of the Pacific by now. We were at home in coral reef waters and palm trees or jungle. We'd seen all the weather including our first typhoon. We knew each other, we knew the ship, we knew our jobs. Secondly: a YMS was such a small part of this incredible fleet. We were one tooth of one cog in this global war machine. Who would pick on us?

I'd only really had one question. How would we stand up in terms of courage? Would we all be brave? Would the Skipper be brave? Would we operate smoothly under severe stress day after day? I knew that my job was clear and I would do it. We all had done our job so far. We had the usual youthful twenty-five year old ego about "My time can't be so soon." We were ready for the Lingayen invasion.

The invasion armada left at two times, the minesweepers a day ahead of the big stuff (battleships, carriers, etc.) and the troop and supply ships. They would catch up.

From the log of 2nd January, the day we left Leyte:

> **0000.** Anchored as before. San Pedro Bay. Tacloban, Leyte, Philippine Islands.
> **0705.** General quarters. Underway from Anchorage. Proceeding to sea in company with Task Group 77.6 for sweeping in connection with amphibious operation in the invasion at Lingayen Gulf, Northern Luzon, Philippine Islands, USS *Hopkins,* fleet guide. Proceeding on various courses and speeds to conform with the harbor and channel to rendezvous.
> **1135.** [Four hours later]. Convoy forming at Point Cay.
> **1225.** Changed course to 183 true. Speed 9.5 knots [straight south].
> **1318.** Took station number one, column on starboard side of fleet guide, seventh in column, distance 300 yards, interval 500 yards [Anti-submarine screen].
> **1450.** Changed course to 218 true.
> **1525.** Changed course to 241 degrees true, 9 knots.
> **1910.** Sounded general quarters. Flash red.
> **1911.** Commenced fire with machine guns only. Two bombs dropped; near misses. They escaped to the south. Ammunition expended. Twenty rounds, .50 caliber. 120 rounds, twenty millimeter. Set condition Two.
> **2118.** Secured port engine due to burnt rocker arm assembly. Cylinder #1.
> **2230.** Port engine back in operation.

2315. Sounded general quarters. Radar report. Distance 12 miles, bearing 355.5 true.
2312. Enemy plane dropped four bombs off our port bow 600 yards, no damage, plane unsighted.
2318. Radar contact lost.
2325. Secured from General Quarters.

And then the next day, 3 January 1945. Let me continue.

0000. Steaming as before. Course 241 degrees true, 9 knots. Set condition Two.
0300. Change course to 265 degrees true.
0420. Sounded General Quarters. Planes on radar, distance nine miles, bearing 005 degrees true.
0430. Lost contact. Secured from General Quarters.
0440. Sounded General Quarters. Many planes on radar bearing 290 degrees true.
0530. Secured from General Quarters. Contact lost.
0710. Changed courses to 292 degrees true.
0718. Speed dropped to 7 knots.
0725. Sounded General Quarters. Commenced firing all our guns at one plane in a group of ten circling and attacking convoy. Four planes seen shot down. One suicide plane dived down starboard side of second tanker in center column.
0800. Secured from general quarters. Set condition Two. Friendly planes in the area. Ammunition expended: Twelve rounds three inch .50, twenty rounds .50 caliber, 420 rounds twenty millimeter, no casualties.
0946. Change course to 330 degrees true.
1630. Sounded General Quarters. Radar, enemy planes 250 degrees true, nineteen miles, CAP, converging toward contact.
1644. Secured from General Quarters. Lost contact.
1845. Darkened ship, set General Quarters.
1950. Secured from evening General Quarters. Set condition Two.
2040. Sounded General Quarters. Radar contact at 270 degrees true, thirteen miles.
2055. Contact lost. Secured from General Quarters.

2100. Change course to 353 degrees true.
2355. Change course to 329 degrees true.

End of two days of log. Do you get the flavor of what was so sparingly noted in these very terse entries in the log? They were typical. We had a five day passage to Lingayen Gulf and this is the way the log read the whole way.

We went south through Surigao Straits, where that famous battle was fought, and then west, north of the island of Mindanao, in what used to be called Mindanao Sea. Then into the Sulu Sea at the western end of the Mindanao Sea. We passed islands quite far off, Bohol, Cebu, Negros. Then northwest by Panay and Mindoro and through the Mindoro Straits out into the South China Sea. We passed Manila Bay, Subic Bay and on to Lingayen Gulf.

It was a five day passage of unrelenting terror. General Quarters, secure from General Quarters, General Quarters. The attacks were day and night. General Quarters became routine. Half the crew slept at their stations, curled up on the deck, heads on their life jackets and helmets under their arms. The helmet and life jacket could go on, the gun ready and loaded in a minute.

That cold, hypersensitive state that I always had when coming into a port or when I had some tight ship handling to do, also flooded over me at the sound of general quarters. My mind raced. We had no team at combat information center (CIC), which all the larger ships had. And on those ships, that is where all the data were correlated. I had to do this alone. And that's the problem with a small ship with one human mind. I have often felt, looking back, that I did not delegate well. I had good officers. Could we have been more of a team? Would it have been better for them? Yes to both questions. But—a huge but—we had dispersed the officers at battle stations, sensible. And there is not a chance that better solutions could have been made by committee!

Meanwhile, the cooks cooked and the engine room was manned and the pharmacist's mate made sick call and the ship's work went on. Small repairs from shrapnel were done but only if necessary. Everyone napped as much as possible. Each day became the next day.

The enemy attacked from the sky on our first night in the Mindanao Sea. It was shocking. The anti-aircraft fire was even more shocking. Every one of a hundred or more ships had many machine guns. All of them firing tracers, from a thousand guns. And they met in a glowing orange umbrella because each tracer line was an arc, like shooting a garden hose. It seemed impenetrable but it was not. And the bombs falling at night were eerie. The pilot couldn't aim at one ship but he sure could see the area of the fleet. He could aim at the tracers. So down came the bombs with that screaming up-doppler and a penetrating whump. Evasive actions had no point. You could turn into a bomb as well as turn away from it. It scared us of course, but any one ship was helpless. This led to a certain fatalism, and the fatalism relaxed us somewhat.

During the day a pilot could aim at a specific ship even if the umbrella of anti-aircraft fire was up there, but it was less visible. Some enemy planes got through and we could not believe that some of our guys in the Combat Air Patrol followed them through. Now that was guts! We saw lots of dog fights, as spectators. When they had shot down the Japanese planes or had driven them out, our guys had to get away, so we let up our firing intuitively, of course.

We had standing orders from the flagship of the task group: "evasive action if necessary." We were alongside a big Liberty Ship. We had no idea of the cargo but it occurred to me, "What if this were an ammo ship?" Down in Manus at Seadler harbor, an ammunition ship had gone up accidently—a fantastic event. No one wanted to be within miles of that. So when we were being attacked I evaded that ship!. We could still do our duty in the section of patrol and shoot at planes and still get the hell away from that Liberty Ship.

All our new guns could fire. How proud we were of our porcupine of a ship. Two long rows of guns, six per side. Some ill gotten goods, but useful. Huge bomb splashes came close. We were actually straddled by bombs. I made a point intuitively, and I think sensibly, of always being on the side of the bridge nearest the danger; I had to see what was coming. It was really the only sensible thing to do, not heroics.

Bomb after bomb. *Kamikaze* after *kamikaze*. We were never hit; we had no burials at sea, not even a cut; now this was luck. But if you were skilled you had greater luck. I didn't know anybody on the crew who was so fatalistic that they used that old saw, "That Bullet Has My Name On It," or "That Mine," or "That Bomb."

My Exec was on the flying bridge and I knew what to expect when the bombs started dropping. He would call down the speaking tube, had a little lid up there; you could hear it click open.

"S-S-S-Sk-Sk-Skipper, I-I'm t-t-too young t-t-t-to die." He tended to stammer at such times and he was joking. The man was wonderful. Usually the cursing of the crew ended with banter. This kind of stuff—gross humor, banter, teasing. It's a great thing to have in battle.

But you can imagine the chilling effect of seeing our Liberty Ship pulling off, the one that was next to us, and they went into Mindoro Harbor, which was at that time being invaded by our forces. She took a *kamikaze* as she entered the harbor and she disintegrated. That was indeed an ammo ship! A huge white cloud rose; filled with thousands of dark puffs, red streaks, awful fireworks in all directions, just sickening. No cursing, no jokes; the word had been around that we had been giving that ship lots of room. The crew knew that. Sober sadness.

The *kamikaze* was an unending terror; we began to see more of them. So much more accurate than bombs; we would call it now a smart bomb. But the sickness of it—the disfigurement of being human. They called it a divine wind but it was stupid; it was a medieval gesture. It was in fact duping bright young men to suicide. Their lives would have been so much more useful to their homeland if they could have gotten back. They were volunteers—hard to imagine. We were all so angered by the damn thing and angered with the country that would do that to their young men. And we cheered when they missed. We cheered when they crashed in the sea. Sometimes it'd just be a big splash but if they were on fire when they went down, there'd be an explosion and the big orange oil fire with black smoke that would finally sputter out.

Some of these planes hit glancing blows and exploded. You couldn't tell what happened until the ship emerged from the smoke; if she had a big black smear on the hull but no jagged hole—then we cheered. Usually there were light casualties when there was only a glancing blow.

We had close *kamikazes*. We only claimed credit for one kill. When one plane zeroes in on you, it comes straight and looms larger and larger. At a distance intolerably close, I hollered "Open fire." Usually this meant the machine guns.

The big gun on the focsle could not shoot at a single plane the same way. It was hand loaded, not a machine gun. We had proximity fuses so we aimed in the path and set up a one gun barrage. The enemy plane had to go through these black ugly bursts. Those three inch shells, which were all a man could lift, made a shattering noise. The ship shook in every corner. Dishes rattled out of the cupboards in the galley, books came out of bookcases. I stuffed my ears with lamb's wool plugs but I have suffered a hearing loss from that time, which has been medically documented.

The machine guns fired in bursts aiming with tracers and when the plane appeared to suck some of the tracers into its body, those were hits.

For all the bullets we fired we had but few hits. Try hitting a butterfly with a garden hose sometime. *Kamikazes* went through that awful flack. Some burst into flames and plunged. But all the flack together must have been some distraction to them, they veered off; they changed targets. We fired at all we could reach; each ship did. A mad splendor.

Radio told us that two destroyers had left Manila Bay and were coming after us. We had one destroyer with us for a guide, the *Bennion* God bless her. We were all such little spit kits! All, but that one tin can, were hopelessly outgunned by the enemy destroyers. Where was our battle group? Where was our support group? I knew the plans called for the jeep carriers and battleships to be up with us on this date. They should be right next to us! Like we, they had been under constant attack. They were a day late and we were naked. So the sweep group was out there alone. The enemy destroyers were coming at high speed. They could stand off and pick us off from beyond our range.

We knew it; the enemy knew it; this straggle of sweeps was not a naval armada! We had that one destroyer as shepherd; she alone had good aircraft radar, alone had guns larger than our three inch .50. And we had yet to run the gantlet off Manila Bay with all those land based *kamikazes*. We were naked, a large target painted on the open sea. We came to know a new kind of boredom, unending terror magnified to unreality by fatigue.

We saw the "Rock," Corregidor. All that history flashed through my soggy mind, but not for long. We had had attack after attack but we did not know how bad an afternoon we would have, a very bad afternoon!

Then came the planes from Manila, probably Clark Field, every one a *kamikaze*. They dived on us from up above. They came in low on the ocean deck, their propellers throwing up water. Evasive action! I went hard to starboard and one barely missed us. Just a few yards from me out on the wing of the bridge. I could see every detail; all of the bullet-riddled plane and the face of the dead pilot and the fire and the smoke. He hit close alongside and I was covered by splash and smoke. There was hollering from below. "Okay, Skipper, okay, skipper?" I waved and called to the helm, "Hard left rudder." We plowed back on course.

I never was able to "chase splashes," the classical textbook evasion when you're dealing with enemy naval shells. There was just too much confusion. And pilot directed *kamikazes* were so different from dumb shells. But I still believed in evasive action, radically at times; hard right, hard left.

Then we noticed our tin can, the *Bennion*. She was firing away at every plane in sight. All of a sudden she did a "180" at full speed. She turned, heeled way over, threw up an enormous wave. There was a sight. She tore after those two enemy destroyers, all guns blazing in full broadside, several miles away. My God, what a sensational event! Full battle, surface, air. The Japanese destroyers were driven back badly damaged and that lovable *Bennion* returned to us in our agony of the *kamikazes*. Planes from the escort carriers way behind us got the word and they finished off the two destroyers with rockets. We heard this on the radio. Cheers all over the ship.

Alone without our destroyer, we suddenly were swarmed by bogeys. Twenty we counted. The LCI next to us (Landing Craft Infantry, about our size) took a *kamikaze* coming in low over the bow. It ripped off the bridge and went over the side. When the ship came through the smoke she was a hull with no superstructure, a creature decapitated. She slowed and turned in circles. Lots of dead guys. One of the other ships took her in tow. We plowed by, still shooting, shooting all the time, as I steered evasive courses.

Our mouths were dry, fuzzy. There was no time to think. Everything was happening at once. There wasn't time to be scared. The attacks just came on. After that big one we still got several planes at a time. And the ships behind us, the fleet, they got it worse. They were bigger and more important targets. The terror truly was endless. Relaxation was a screaming need, but the naps were short.

Looking back I was especially driven by an intuitive drive to protect my family. And that family was the crew. I think this is an intuitive human drive, a biological drive. No room for fear. Fear cannot take precedence over instant action. Things had to be done. Were we heros? I didn't ask that question then. But if you ever put the ship's survival above your own you're a hero and that was sensible and it was intuitive and it was training. There were no flashing dreams of heroism. It was just doing unthinking skills one at a time. In that sense heroism was everywhere and mundane.

Going into Lingayen Gulf I was satisfied with some of the things in the previous four days. Those extra guns we stole were priceless because we put up a wall of bullets. Hit anything or not. Twenty-four guys out of the thirty were working guns. They were involved, not standing around. The number of rounds of machine gun fire was not usually noted in the log, but they were extraordinary!

The gun crew discipline was very good. These were just guys off the street now in the Navy but their gunnery discipline was terrific. No one fired before I hollered from the wing of the bridge, "Open fire!" They just had to lean on their guns and wait. That was tough. What trust! We had a talker on the bridge, a guy with a great big helmet, outsized helmet with ear phones and a telephone. And he passed the word to fire by phone to talkers on the fantail, way aft, and on the focsle, by the three inch .50 gun. Two of the officers had local control of machine gun stations.

I felt, by the time we got to Lingayen Gulf, that I was a veteran. I had made no grievous mistakes. And if I had grown in my job, so too had all hands. The citizen sailors were now salty. They knew their duty. Every one of them. This was surely our worst of times and we had made it, all of us.

But it was not our worst of times.

We sweeps entered Lingayen Gulf with early morning mists still clinging to the water, a stunning panorama. This was on the 6th of January, 1945. It was to be called by naval historians that I have read since, "the red rain over Lingayen," and "one hell of a day." "A gruesome day." Read this verbatim page from the log of S-Day minus three. We had three days of minesweeping before the invasion, the usual for minesweeps.

> **0000.** Steaming as before, course 006 true, speed 9 knots. Sounded General Quarters, radar contact 050 degrees true, distance 17 miles.
> **0215.** Lost contact, secured from General Quarters.
> **0220.** Change course to 051 degrees true.
> **0233.** Sounded General Quarters. Radar contact, 045 degrees, distance 7 miles.
> **0310.** Lost contact, secured from General Quarters. Change course to 093 degrees true.
> **0520.** Sounded General Quarters. Radar contact.
> **0530.** Secured from General Quarters.
> **0543.** Sounded General Quarters. Enemy plane dived near the column of Unit 8. No casualties.
> **0658.** Orders from CTG 77.6, 'Proceed in accordance with basic plan.'
> **0710.** Proceeding into Lingayen Gulf, Luzon, Philippine Islands with Unit 7.

0745. Change course to 160 true.
0748. Enemy planes attacked Task Group. Ammunition expended: three rounds, three inch .50, plus AA.
0800. Steaming into Gulf, following AM unit, who were sweeping a channel. Report 58% fuel, speed, 10 knots.
1000. Change speed to 9 knots.
1129. Enemy planes in our area being fired on by all ships in Task Group. General Quarters.
1207. Secured from General Quarters.
1255. Two enemy bombers approached from starboard at mast height. One crashed DMS in the bridge and the other hit APD aft of bridge. Both vessels on fire.
1315. Proceeding slowly with Unit 7.
1400. Ordered to act as mine destruction vessel, Unit 7.
1500. Unit 7 begins sweeping area "Shackle."
1710. Ceased sweeping area Shackle. Ships began recovering gear.
1755. Enemy planes attacked fire support group which were standing into the Gulf 10 miles out from us. Observed enemy planes making suicide dives. Hit three ships, missed two.
1825. All enemy aircraft out of area. Ammunition expended three rounds, 3 inch .50, no casualties. Secured General Quarters.
1837. Sounded General Quarters. One enemy bomber again suicide dived into bridge of DMS on our port beam.
1845. Four Black Widow fighters in area for night cover.
1930. Change course to 345 degrees true. Night retirement from Gulf. Second in column of Unit 7, speed 4 knots.
2000. Secured from General Quarters.

That log from the first day in the Gulf. There were six more days like that before we were to be sent back to Leyte as escort for a convoy of empty LST's. We swept for three days before the invasion; we got hell each day. Completely exposed to land based air power. I recall episodes from that time. Log entries have confirmed them. We were scared. We were keyed up. We were exhausted. Impressions were exaggerated, sharpened and blurred. Life was a dream being lived.

The Gulf did not seem to be more than a dent in the shoreline. Actually it was about twenty miles across, thirty miles long. But in the center it hardly seemed like a bay. Huge. Green jungle covered the mountains all around. Heading in, there was an island to starboard, Santiago Island, and it seemed to have old cones for mountains. Not very big, extinct volcanoes. The end shore of the Gulf looked like a coconut plantation. This was part of the low plain which led south to Manila, a hundred miles away. Our troops could use that to take Manilla, exactly the route of our enemy three years earlier.

We were sweeping the sky with field glasses all the time, looking for planes. It was a beautiful day. White clouds, good for planes to hide in. The water of the bay was a lighter blue; it was less than fifty fathoms deep. Spectacular anchorage if you were interested. The shoal waters were that blue green of a tropical lagoon with white sand bottom.

Out on this beautiful picture, a clutter of minesweepers was forming into echelons. They were sweeping lanes to the shore. It seemed like a play that we were watching, or like a game with model ships. This was 6 January 1945. Three days before the invasion. The planners had dictated the entry into the Gulf at 0700. Our log said we were ordered into the Gulf at 0658, two minutes early! Two thousand miles away on that huge slope up from Lake Sentani in New Guinea mountains, those Quonset huts had inhabitants that decided all of this. And at 0658 orders from CTG 77.6, "Proceed in accordance with basic plan." Incredible! Outrageous, sublime planning and execution. Our world was like a huge corporation, like General Motors. It was a giant corporation planning the war machine. Design and production of the war machine, the entire nation making war. The communications and the integration to make war, all the mimeograph machines, all the paper. It worked. Being part of that machine leaves no room for whim, no personal likes or dislikes. In battle we were one cog on that one wheel within that huge machine. The wheel turned, we turned. We did what was needed now. No introspection. That came between battles, I can assure you.

The *kamikazes* came down from the clouds. The log noted that two bombers hit two destroyers near us. They came at us just off the water. We were headed into the bay to sweep. They came at us from the starboard side. Coming out of the bay were two destroyer minesweepers, one on each side of us. They had made a fast run in and out, clearing a channel for the rest of us to widen. And this was when the two planes, the medium bombers, came at us, when we had a destroyer exactly on each side. The starboard tin can got hit on the bridge; the first one, on our right. Huge explosion, fire, men running to the focsle fleeing the flames. Carnage.

The second plane flew up and over that burning can and headed for us. Right for me!. My eyes were glued; and I knew that in seconds I would be dead. The plane loomed huge. As it roared the two-three hundred yards from the exploding tin can which now struggled past us, our twenty millimeter tracers plowed into the plane. Too late. Then I ducked behind the canvas spray shield. At the last second I ducked and stared at that plane. But it pulled up. It climbed up and over us. The pilot had seen the other destroyer on our port side. It just cleared our mast and socked into the engine room of that other can. Two hits out of two destroyers. They were on fire, and in sinking condition.

We immediately knew what saved us: two larger targets. We were in formation, in line, moving at nightmare pace. We were a mine destruction vessel, no sweep gear out. My first reaction was pull over, fight fires, help those poor guys off. I did not do it; I followed orders. Stayed in formation. An Australian cruiser and a tug had come up, and of course, they did a much better job of putting out the fire than we could have, but I wondered for so long if I had been a coward. That was tough and it nagged and it didn't go away for a long time. It happened so fast. It all happened in seconds.

There was chaos on both those destroyers, with the fire and explosions. They lost speed and settled; one was abandoned, anchored. Smoke rose high in the sky; it was a sudden nightmare, impossible. The other victim was towed away.

We slowly steamed in at four knots and I kept remembering the charming group of officers that had us to dinner on that ship, now dead. How many of them survived? Such nice guys, each one a vivid character. And what happened to the monkey? Unreal question? No.

The cook brought sandwiches and coffee to all hands on the ship. We had not sat down to a meal for days. The working deck gang manned brooms and swept the shrapnel from our littered decks. There were lumps of men curled around life jackets in the sun on the deck. No coffee for them. They shut out the danger in sleep between attacks.

The bay was a huge mural, the water and the sky so blue; the border so green. And on the blue were gray toy ships slowly moving in staggered lines toward the shore, minesweepers. Then the whole gang of toys would become real ships when bogeys dived on them. The air filled with dirty puffs and orange tracers and some of the gray toys stopped and burned. Not toys at all. Real guys. Toys don't have people smashed and burned; toys don't have men literally postponing their deaths to save their ships; many such cases. Only the gray toys of our own formation were life sized, and we were shooting guns and sweeping area shackle.

The day did not end; it didn't end it just turned into tomorrow at 2400. And we swept for magnetic mines and for moored mines. Different gear, but gear that handcuffed us as fighting ships. No evasive action, we were slow targets (four knots), dragging hundreds of yards of cables, floats and depressors. All this in slow motion like the nightmare of running in sand or swimming in syrup. And all the terror couldn't speed the ship. It was agony ending in bone-breaking exhaustion.

One afternoon I called for my Executive Officer and quietly told him on the bridge that I was about to pass out. It had been a little bit more than three days without even a nap. And he grinned and said, "I relieve you, Sir." He pushed me down to the deck, shoved a life jacket under my head. I looked up and there were two of him and the bridge was slowly rotating like a merry-go-round. It was time for me.

Suddenly I leaped to my feet and the Exec grinned at me again. "Well, that was a nice hour's nap."

"My God, I'm glad there was no air attack."

"But there was! There were five."

"No! Did we fire?"

"Hell, yes! Every gun."

My vision was clear. So was my mind. Felt wonderful. Good thing. There was more to come, but with a buddy like that.

Our unit recovered sweep gear to retire for the night on the open sea. It was safer out there. On this "night retirement" we would go slowly out to sea and then come back at dawn. But the bosun didn't like the way our gear was acting while we reeled it in. I went down on the deck and looked at it with him. It was plunging at an odd angle; it could be a hung-up mine. I hollered up to the bridge, "Call the unit commander. Tell him we're going to try to shake loose a possible fouled mine."

The rest of the unit pulled away and we started radical turns trying to snap our cable, loosen the mine. Finally we decided we were clear and we recovered our gear. Our unit by now had hauled ass. They were long gone and the 353 was the only ship in that huge gulf at sunset. Not quite alone. The burned out hulk of the tin can, hit earlier that day, was anchored and still smoking. No one was aboard. She was very low in the water, so close, a dead ship. She was a grey ghost over there. And we all remarked that she looked like she was steaming along slowly. Wisps of smoke still came from her crumpled stack; and there was a little wake astern as the incoming tide kind of roiled by; left the bubbles. Having her for our only company was macabre.

Then, "Bogey! Bogey! Bogey!" High in the clouds above us was a single plane. We couldn't even see what it was, but it was a bogey. We were already at General Quarters for sunset. He peeled off until he had us dead aim. No pun intended. He came in a power dive. We were holding off to fire everything, ready to put up a cloud of lead. I held off until we were sure to get him in one blast, our only chance. I swung the ship around to put him astern so that every gun could bear. Closer and closer. And then, all of a sudden, just before I could holler "Open Fire!" he swerved! At the last moment. And he heeled over in a sharp bank and snapped into the stack of that abandoned destroyer. Right next to us. Blam! Fire and smoke! And she was in death throes again, finally. I saw everything in exaggerated slow motion, and we were all silent, slumped. Then a cheer. And then shouts and cursing.

I was out on the port side of the bridge nearest the destroyer, just forward and up from the twenty millimeter gun crew. The gunner unstrapped himself from the harness and he kneeled on the deck and he roared, "You unlucky son of a bitch, you unlucky son of a bitch, you unlucky son of a bitch!" He was laughing in a way that sounded like crying. We went to flank speed and tore out of the gulf. Now, truly alone on that vast bay.

That was the most magnificent sunset I've ever seen, before or after, anywhere, and we were accustomed to the special sunsets on the South China Sea. Massive. It was seen through eyes which still worked, thank heavens, not eternally dimmed. It was a mackerel sky over the entire dome, all sides. Each tiny fleck of cloud turned into a golden coin, perhaps from the Spanish ships that once held these waters. All over gold, not just the West; the entire sky! No one spoke. It was unbearable. It was overpowering. And suddenly gone. And the blue of the sky and the sea faded to black, and we had to find our fleeing friends by radar. Four Black Widow night fighters flew low over us in formation—we saw the dip of the wing saying, "We have you covered."

When we found our unit, it was in two columns of four YMS's. We moseyed along at four knots straight out to sea. Suddenly we relaxed; many men got a real sleep. We had some Red Alerts, General Quarters, but we were not attacked. And the sky full of stars seemed so close and so unaware of us and our unending terror. Unreal feelings again. And I said "I'm going to try my bunk; anything comes up, call me." The slow churning of the main engines was my lullaby.

Then the voice tube and a sharp voice from the bridge. "Better get up right away, Skipper."

"Okay."

"Look at the radar," they told me when I got up there, "what a mess." Our little group of eight sweeps was heading directly for the center of one of our large attack forces on their way to a dawn bombardment. Looked like big stuff; huge blips on the screen. Several YMS's called the senior officer of our group, asked, "Are you gonna turn, are you gonna turn?" No answer, nothing. One captain said he was turning off; he just all of a sudden said he'd turn. We saw a blip veer off on radar. He was on the outside, last in line; he could make it, but it was too late for most of us. I decided that turning would be terrible, put us broadside. And they were roaring at us. The distance closed. Suddenly they were much less than a mile away. I couldn't see them; inky blackness. But we could hear the voice now on the radio, so loud, so close. "Chickens get the hell out of the way! Get the hell out of the way!" What were they talking about? It was too late! Not enough time to turn.

We immediately sounded General Quarters. The crash would be so soon. We passed the word, "Collision Station." They came at us in several columns, several to a column. We couldn't count them on the radar they were so close together. Suddenly we could see them; suddenly we could see the bow waves, and then the actual bows of the ships themselves. They were coming straight at us, shadows in the blackness. I hollered in to the throttle man on the bridge, "Turn on all the running lights! Now, now!" Suddenly we had two white range lights forward and our red and green lights on each side of the bridge. First time I had ever seen them on any ship I was ever on. They worked! Lights were an unforgivable sin during war.

By now the bow waves and the hulls of some cruisers were dead ahead. We could see the white bow waves and they did not slow down. Then the lead cruiser suddenly turned on her lights. Then all over that task force, lights! All the YMS's put their lights on. There were frantic course adjustments. Huge looming shapes came out of the darkness and materialized all at once. One of them just missed us. Lights everywhere! The radio had many bad words. Suddenly we threaded through their whole column. They were astern. Lights went out. Blackness returned. Unreal feelings again.

I did not get back to sleep that night! There were coffee and sandwiches from the galley. Boy, the stories, next day. "We missed by twenty feet!" "Ah, ten!" So good to be alive. You know, death was averted again. Damned feather merchants turning on lights!

We were living the war, not pondering its meaning. War was death—death for the enemy and for us. All our tools, skills, and attitudes were honed for inflicting death and averting it. It was that simple.

For most of us our reactions and our outlooks were not related to religion. It was said then that "there were no atheists in a foxhole" and I think that was garbage for most of us. I don't think religion had a great deal to do with it for most of the men, but of course for many there was a sustaining faith from childhood. Prayers were said of course and some grasping of religious symbols. But I think this was not a towering divine conviction; this was more like rubbing rabbit's feet, ritualistic. All of our minds became mangled in some way, but war had become routine.

Next day we formed up, swept for moored mines. Our job was to be last in line and shoot up the mines that were cut by the ships in front of us, so-called "mine destruction vessel." Then all of a sudden on the radio it sounded like the skipper in the number three spot was having troubles. And from ComMine, "Assume Station Immediately," and from that ship, "Unable to do so, Sir."

"Fall out of formation. Mine destruction vessel assume number three position." We tore up ahead; we had the gear ready and were ready to stream it as soon as we got there. Everything was set up before hand. So a captain had personal problems. I knew the guy; nice, quiet. But his ship could not do its duty. This was a real failure and I was upset by it, worried about my friend. He was quickly relieved of command and shipped home.

Finally we saw our fire support group on the horizon, coming in. By God it was high time. But what a sight! This was raw naval power. This was calendar art. The charging of the battleships was classical; never seen since. They caught up with us a day late. Our day of hell. And the carriers sent more Combat Air Patrol. Ah, such a brave sight! There were four old battlewagons in column. They were the ones that were sunk at Pearl Harbor, had been repaired and were now used as a bombardment force. They weren't fast, they didn't go with the carrier fleets. The new battleships did that. But these old wagons plowed the water. They were wide, wider than they were tall, and they would squat down in the water. They were ugly; so impressive. Nothing could stop this might, so heavily armored. Inches of steel in their skin.

Behind the "battle wagons," were cruisers, and all around, destroyers sort of dancing. Outsized flags snapped. Each wagon settled down behind a huge bow wave they plowed up. And the wake of one ship bubbled into the bow wave of the next one, which formed a foaming highway for these heroic sized ships.

The *kamikazes* were still around. Some missed, some bounced off the heavy plates, but some hit towers and some hit gun mounts. There was chaos in the sky. We were almost used to it now. The big guys were so much more able to take on the planes than we were. But even the huge bulks of wagons were slowed and some were stopped. Diving planes, explosions. So many casualties I found out later. The planes with the dreaded meatball ruled the sky. The question, "Would we have more ships than they had planes?" The mighty armada was clobbered. There were dreadfully damaged ships. A ship might veer off to fight the fire but did not slow the rest.

They plowed on. And then in the middle of the Gulf they opened fire, even while doing damage control. My God, shells over a foot in diameter weighing almost a ton. Cruisers joined in, and destroyers with their five inch stuff. All high explosive. They drenched the beach in a creeping barrage, moving inland. They were here to protect us and they did. They protected us by being larger, more important targets to the *kamikazes*.

On the first day of sweeping, twenty-one ships were hit, some of them more than once. Most of them stayed afloat. Casualties were terrible.

There was a fundamental difference between U. S. and Japanese postures on manhood somehow. We frequently used weapons in overkill, like this very bombardment at Lingayen Gulf. It was not needed but we did it. The U. S. posture was "bullets not boys." We expended money and steel and we saved lives. But the Japanese differed. They saw the nobility of mass attacks, even failures. Their planes were bird-like and graceful. There was no armor for the pilot; they were light and maneuverable. Ours were sturdy, with armored cockpits, self-sealing gas tanks, slower and stubby, but the pilots were saved. And our hospital ships were right behind. The *Hope*, the *Mercy*, the *Solace*.

Did I hate the enemy? Yes! We hated constantly and deeply. We hated everything Japanese from the Emperor to the macabre pilot diving on us. Of course the pilot was also the target of our bullets. The hate did not gnaw at me personally; it was a pervasive every day feeling. In our daily mundane lives we did not dwell on the ultimate madness that drove us. That madness was our line of work: war.

Even before S-Day we anchored in the Gulf at night and we were suffocated with that smoke just like Leyte Gulf. Those smoke pots on landing craft would move among us spreading those stinking fumes. Good night. Taps. *Kamikazes* did not dive into miles of smoke. But reports came in about suicide boats carrying torpedoes or bombs. Swimmers had bombs strapped to their bodies, we heard. Most importantly these robbed us of sleep, which we just craved. I never saw one even though we were up all night running around searching with glasses. We carried rifles and pistols. Never fired them. In the morning we saw no damaged ships.

Finally the three days of sweeping was over. Sugar Day came and it was completely successful, the most effective ever, to date, in the Pacific. We finally controlled the air. We, the sweeps, had spent days, absorbing what the troop transports were spared and we couldn't argue with that. The night before S-Day we did laundry; hadn't done it for days and days. And invasion morning came; the sun rose, and we had clothes hanging out all over. I can imagine the cruiser captains with their glasses on us. "Not very Navy," they must have fumed.

Again came the bombardment down the gut of the Gulf. The big stuff again but even more this time. What a bombardment! The fleet carriers had sent in wave after wave of planes. They came in formation and they peeled off for attacks, just like in the movies. They bombed airfields on Luzon and the beaches and the roads to Lingayen.

Our ships were miles offshore. We could see the flashes but not hear the thunder till much later. The shells came over our heads and in the distance the sound was continuous like rolling thunder of a summer storm. And the smaller shells screamed over us and went "crump" into the palm trees. But the battleship shells, they rumbled, honest to God, like a freight train going over a lousy set of rails. They rumbled! You could see each of them! Right overhead. One at a time you could see them and count them. And they went "KA-WHUMP!" The sky was full of destruction but it was safely over our heads. Must have been thousands of rounds! Unbelievable millions of dollars. Nothing could make a human feel more like a flea. It was intimidating.

Then in a long line abreast came LCI's and LSM's, scores of them. These were ocean going amphibious craft but they were covered with rockets pointing forward on angled racks from which they were launched. About a mile off shore they stopped, lined up, and they fired almost simultaneously all along the beach. They went "Zoom-zoom, flash, flash." They were rockets and rose in a parabola to reach the beach, waves of these things. They went up, arched, and plunged in a sort of moving cloud. We could see all these things as a mass of rockets. Then huge clouds of dust and smoke, the shores were churned. I imagine there was no town of Lingayen left.

Finally our turn. The laundry was taken down, everything taut. We were to make one last sweep in as shallow water as possible. And we wheeled in and out. Our gear was behaving perfectly, very shallow setting to go into very shallow water. We all did it smartly; we were veterans. And nobody shot at us; one YMS was damaged by shell fire from somewhere on shore.

We retired from our sweep to our own anchorage again and we ran into uncountable small landing craft. They were circling around their troop transports, before turning towards shore. They were waiting for that flag hoist out on the command ship to be "two-blocked" to the top of the mast. Up it would go; then down it would go and off they would go to the beach.

LCI's followed with troops and then the huge lumbering LST's; these all could go right onto the sand. They advanced in a long line abreast right to the beach. LST's carried floating ramps to make instant piers. No enemy artillery, only occasional bogeys. But by now the Combat Air Patrol would be driving them off and we ignored them. The landings were textbook, the plot more complex than any theater. And "a cast of hundreds of thousands."

Our job was done. The Gulf was a churning mass of a thousand ships and more thousands of small amphibious craft all leaving little white wakes behind them.

Unreality set in again. Our exaggerated senses again. We watched this game of tiny ships on a game board. I was alone in my thoughts. It is not an empty phrase that the captain has the loneliness of command. And I mused; we had done the job. But you can only muse after the job is done. In the wardroom we were buddies. Somehow we survived all kinds of danger in the past few months, the moods of the sea and the personal tensions. We all made it and the weariness and the boredom, even of combat. We'd had more *kamikazes* than anyone predicted, but we'd had almost no response from the shore in terms of artillery and almost no bombing. And almost no mines! All that sweeping. Sixty sweepers, three days. How come, almost no mines? I want to quote Morrison in his Volume XIII of the 1959 *History of U.S. Naval Operations in World War II*. Page 112, he said

> Admiral Kinkaid's action report has a story of Filipinos with bancas who, having recovered in Lingayen Gulf some 350 mines, took them apart to provide themselves with explosives. Nobody now believes this yarn which was probably told to the Admiral in order to explain the faulty intelligence estimate of five mine fields in the Gulf.

We were astonished that these waters had no mines. We knew nothing of the Kinkaid report. We shed no tears. We gave three cheers.

Then suddenly, two days after the invasion we got orders along with some other YMS's. At 1547 that afternoon:

> YMS's added schedule for slow convoy departing from vicinity Point Charlie at 1830 tonight following APA Convoy commander 98 in Sherwood OTC and CTU 77.6. A senior YMS commander ABD will ensure prompt arrival all ships at rendezvous point.

We were headed back for Leyte retracing our passage of a week or so before. We still were attacked, the terror was still unending but it certainly was diminished to a dull ache.

Twenty years after that minesweeping of Lingayen Gulf I was on a plane from Clinton, Iowa to Chicago. I sat next to a very fine gentleman and we conversed for an hour. We parted; I don't remember his name. I had introduced myself as a professor at the University of Iowa and he was a retired Army general.

"Where were you stationed?" I asked.

"In the Philippines."

"How long?"

"Four years. Never captured."

I said, "Somewhere our paths must have crossed. How?" Well, he had worked just east of Lingayen Gulf in the mountains. He built a Filipino Army. I asked him the question: "We could find almost no mines in Lingayen Gulf. Can you tell me why?"

He quickly answered, "We took them all out. Hundreds. Not I, but the local fishermen who worked with us. We organized it." They would see them from their bancas in the clear waters while they were fishing. They knew those waters so well. They put out a buoy. After dark they returned and unshackled the mines from the cable; they hauled the mines back very carefully, took them apart for the explosive. They got used to it; they knew what to do. The general said that he took half the explosive for his army and the fishermen had the rest.

I was stunned by this; grinning I turned in my seat and said, "Sir, may I shake the hand of a man who probably saved my life."

Subic Bay: The Liberators

WESTERN LUZON, PHILIPPINE ISLANDS

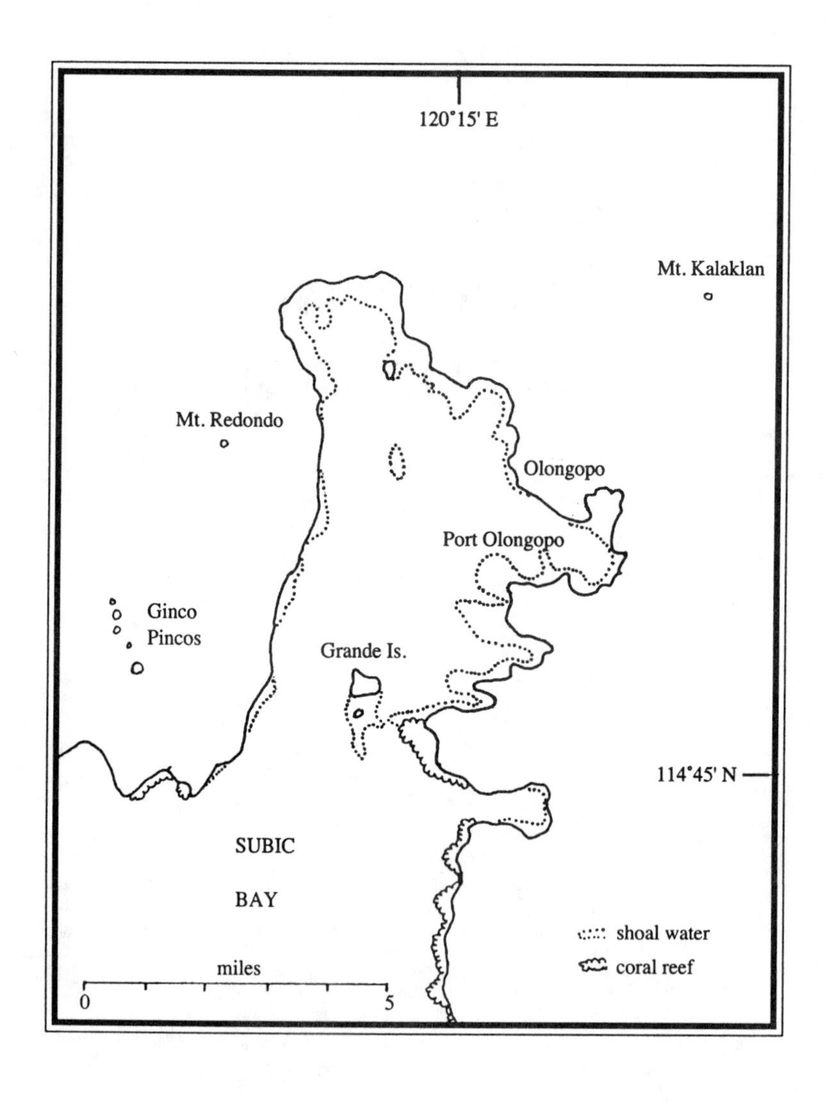

There were only a few major invasions in the Pacific war. There were so very many modest to minor ones. The log of 20 January, 1945:

> **0000.** Steaming as before on various courses and speeds. Maintaining station on anti-sub screen off San Antonio beachhead.
> **0605.** Underway with minesweeping units and escorts to Subic Bay, course 130 true, speed 9 knots.
> **0820.** Fuel report—50% capacity.
> **0930.** Lying to off Subic Bay preparatory to sweeping western half of area. YMS 353 to serve as mine destruction vessel.

Those very brief entries record two invasions after Lingayen Gulf. This was an incredible anti-climax to Lingayen Gulf which for days went on and on; incessant air attack, threat of mines, shore batteries. One of the greatest invasions ever mounted; a terrifying, horrifying period of time.

After a run back to Leyte, escorting empty troop ships and LST's, we returned to Luzon to land troops at San Antonio and at Subic Bay to cut off the Bataan Peninsula to the Japanese. These little invasions were just south of Lingayen. This was very small scale by comparison. We did have cruisers and destroyers, cargo and troop ships and of course, a dozen YMS's or so. Always the YMS's. We escorted them up from Leyte; then we swept and did anti-sub patrols. Landing the troops at San Antonio was against no opposition but the Subic Bay venture was to recapture an old U. S. Navy Yard from the days of our bases in the Philippines. Surely the enemy would defend Subic Bay.

This gorgeous, and certainly one of the best harbors in the whole world, had been heavily fortified by the U. S., and mined. The Japanese took it over at the end of 1941. Grande Island blocks the entrance of the bay. It was fortified. And the single channel by the island is less than a mile wide. The narrow channel could be covered by the heavy guns on Grande Island.

Just off Grande Island we were lying to for the last softening of the island and the approaches to the bay. This little island, had never been subjected to anything like this. The cruisers and destroyers were standing by several miles off-shore. But from the sky the aircraft carrier planes were dive-bombing it. The fighters were coming in, strafing. It was one huge pall of dust. There was nothing, nothing there!

And then, all of a sudden, the strafing and the bombing stopped; we were ready to go in and sweep the bay. We had just started in, when this small Filipino man in tattered clothes (I watched him through the glasses) climbed from the rubble of Grande Island and walked into the channel, in shallow water. There he waved a small American flag, obviously home-made flag; looked like it had been made on a pillowcase with crayons. I was close enough to get a very good look at that terribly frightened small man waving that flag so forlornly. He had seen our American flags. A small boat from one of the LST's picked him up. I don't know whatever happened to the man. How he survived that bombing I don't know. He was the only person there; no Japanese on the island. This was what the United States did in World War II wherever they went; they pulverized the beaches. Overkill, and not always completely effective.

Subic Bay was incredibly beautiful. It was ringed by green, tropical mountains. They were so steep and so dense you could see nothing inside. The trees and vines and the epiphytes (the little plants that grow on the trees' branches) formed a blanket. The solid wall of forest plunged right down to the water. And the water was a stunning clear blue with coral along the edge, a fringing reef. And that was the very, very pale blue around the dark patches of living coral. The white sand showed pale blue in the water. We were used to it by now. Saw it all over.

We were sweeping in echelon, half a dozen of us. The lead ship was farthest to the right and we were the mine destruction vessel, last in line and close to shore; we were also the only covering support. In other words we had all our guns trained on the jungle, manned and ready, in case any shore battery should open up. The destroyers couldn't get in there until we swept, so we played destroyer. None of the other sweeps had our fire power!

Tension was palpable. We were so close, only two—three hundred yards off the beach. The bottom sloped down steeply there; the charts were good on this. We knew good water by now in the coral areas; we could get quite close. But the nakedness of this ship flying that American flag up there so bravely was scary. We had no idea how many Japanese guns were trained on us. We could have been subjected to rifle fire. An anti-tank gun could have cut us to ribbons. It was terribly tense. The machine gunners were strapped in a harness, and they leaned on the harness, hunched over.

We tried to make ourselves small. We wore these huge kapok life jackets which weren't all that good in the water after a while, but they were marvelously thick, about three inches. We felt that they would stop a bullet somehow. They wouldn't, but we felt secure in that thing with the big high collar around the neck. We pulled the helmet tight and hunkered down.

I was on the shoreward side of the bridge staring with binoculars, as were other officers and the lookouts as well. We came on a clearing with houses. It was a village, oh, a hundred yards from the shore, up on a little grassy slope in this otherwise precipitous mountainside. There were no people. It was dead. I thought, "Oh, good Lord, they have gotten rid of the residents and they've put guns in those Nipa huts and all of a sudden the wall is going to be knocked down and we'll be looking up at anti-tank guns." Because they certainly, certainly expected an invasion and we gave it away with all our bombing and the massing of our ships offshore. When we saw this empty place we were just absolutely looking for the worst. We were targets. Average heart beat? High!

Then all of a sudden there was an eruption. Not an eruption of gunfire but an eruption of people. Suddenly there were all these people. They came out from the woods. They'd been hiding; they didn't know whether we were Japanese or not. Then they had seen that fluttering Red, White and Blue on ship after ship of the minesweeper group as we swept steadily by, searching for mines.

The villagers were laughing and cheering and they were hugging each other. With the field glasses I felt like I was there with them in the yard. Four old men came down to the shore; they made their way slowly. Then drew themselves to as much full attention as their tired bodies would allow. They were frail old men. And they saluted and stood at salute while we went by. They were honoring the United States flag. We had returned, as promised.

The women held hands and danced. They had several little round dances. I didn't know anything about their dancing, but this was obviously dancing for joy. And I couldn't hear the sounds they were making, but they were certainly laughing and shouting. Four little boys pre-teens, striplings, slender, skinny kids, jumped in a canoe and paddled out to us. They caught us easily, we were going so slowly. And they pulled up alongside. And this one little spokesman, skinny kid, very few clothes on, stood on the outrigger with a hand on the backstay. He cupped one hand to his mouth, and with a face filled with this huge, beautiful grin, said so clearly in a steady voice, "We are so virry hoppy you have came back." We just collapsed; not a dry eye on the ship.

Then I called to him, "Where are the Japanese?"

"They ran when they saw you coming. They all gone. Our men are chasing them." The boys wanted to come aboard and go with us. I couldn't let them.

We absolutely lost our bearings. Suddenly from the extraordinary tension, from this terrible bombing right next to us on that island, and then going in loaded with fears; suddenly there's joy. The men slumped on their guns. We could hardly stand up. The relief was physically almost unbearable. The new experience sank in; we were liberators. This was a hard thing to understand, and a rare event for sailors to be close enough to talk with the liberated people.

I passed the word to beware of tricks. We of course stayed at battle stations. By the way, only one mine was cut I think, and possibly that was one of our own from years ago. We still looked in the green depths of the jungle, still looked for gun flashes. There were none. Just laughter and tears and dancing.

However, when we got to where we could see Olangapo and the Subic Bay Naval Base, they had been torched. Black smoke poured up from orange flames. There was no city. This was beautiful Subic Bay. We had returned it to it. Shore facilities could be rebuilt in time, and ours was the first step.

We kept up the minesweeping, but the words "We are so virry hoppy you have came back" hung in the air, forever etched in our minds.

I wanted to relive the history of Subic Bay where our Navy fought so desperately in December, 1941. This was part of the Bataan Peninsula of such infamous repute. I had a copy of a book with me, *They Were Expendable,* the first story of PT boats. This was in the very short time before Corregidor fell to the Japanese in early 1942. This book was written by W. L. White about a young lieutenant named Bulkely, who was the commander of this little group of half a dozen little PT boats there at Corregidor (actually in nearby Mariveles Bay.) It was a lurid account of the heroics of these PT boats. The Japanese had taken Subic Bay as their naval center. And it was close enough to Manila Bay and Corregidor that the PT boats could attack and return to the base in darkness. The book has a long list of such attacks and a long list of the sinkings and great victories. They sank and disabled, ship after ship; cruisers, troop transports, destroyers, tankers. For many of the ships they gave the exact place within the Bay where they sank.

I hauled out the book and plotted all these sinkings on our chart; we had very good charts of Subic Bay. And one afternoon in our four day stay there, I rowed over in our little rowboat, to the sites where all these ships had been sunk. The water was very clear and I could see the bottom, the coral patches all so vivid, and the schools of reef fishes casting shadows on the patches of white sandy bottom in between the coral. It was beautiful visibility, not even a hundred feet deep in most places, and in the bays even more shallow.

I did not find one wreck, not even one piece of a wreck! The book was a sham on this score. But it sold like gold!

Heroic men? They certainly were. Night after night, and they got their boats shot out from under them; sailors died. But effective they were not. In fact Morrison in his post-war series on the history of the Navy in World War II says that the PT boats sank almost no ships, anywhere, and certainly not here. In fact Bulkely, in his book on the PT boats after the war, then a Captain, said, that there were no sinkings in Subic Bay. There were a few ships damaged and many barges were sunk. But that's not the story that came out in 1942, which in fact made a great John Wayne movie: I've seen it quite recently on late night television.

We were only in Subic Bay for a few days. We did routine sweeping, the entire bay. We did anti-submarine patrol off Grande Island. We refueled from a large minesweeper. We scrounged for supplies. We loafed. One evening, at the usual General Quarters, a radio message came over: "Friendly medium bomber reconnaissance plane in area. DO NOT OPEN FIRE."

I hollered down from the bridge, "Friendly planes coming. Hold your fire. Hold your fire." The horror of friendly fire; we had seen enough of it. Then a destroyer opened up on a plane. The plane was a medium bomber. And when it went over the destroyer, they opened fire. Then other ships down the line. So much for getting the message about this thing. Our men were tracking with their guns. I hollered again, "Don't fire, don't fire!" It flashed overhead. Big red meatballs on the wings! "Open fire!" All of our guns opened at once. The men were straining in their harnesses. It got away, probably wondering at our laxity.

And on the faces turned up at me from the guns was a reproach that I'd not seen before. It was just so awful to have an enemy plane fly free; I mean this guy was flying low! It was obviously a Japanese reconnaissance. They flew right over us at low altitude only getting following fire; all missed. I passed the word to the helmsman, "For heaven's sake pass the word about that radio message." The word was passed; fewer dirty looks after that.

Next afternoon we were "Anchored as before, one mile east of Grande Island, Subic Bay, Philippine Islands." That's from the log. At 1644, the log says: "Underway from anchorage. Proceeding to sea as escort of convoy en route to Leyte, PI." At 1725, not an hour later, "Course 209, speed 8 knots." We were at sea, in formation. That was a frantic hour!

It was a hot day, ninety-five degrees. By late in the afternoon the ship's work was done. We had been admiring the precision drill of a convoy leaving the bay. Destroyers dashed out first, then a couple of large, empty supply ships and transports, plus some minesweepers. This was when a messenger dashed up to me and said I was wanted on the bridge right away. We had a radio message. The radio message in effect was, "Lena Horne 353, underway immediately, join convoy screen, starboard beam of flagship."

We sounded the blast on the alarm, "Special sea detail! All hands on the double! Engine room, fire 'em up. Let me know when we can haul. Bosun, up anchor on the double." In seconds the helmsman was at the helm, black gang was in the engine room. There were engine rumblings down there. Radarman, signalman on the bridge; so on down the line. We passed the word that we were joining the outbound convoy. Everybody was dropping whatever they were doing, running to their stations. The rumble of the engines was followed by belching black smoke from the stack. All loose gear was being stowed. We were ship-shape in minutes. "In all respects ready for sea."

Then a call from the engine room, "Ready for half-speed. What the hell's going on anyhow?" They were always in the dark down there.

The bosun cupped his hands and called up to the bridge, "Anchor in sight."

"Engine room, how soon for full speed?" I called down.

"If you need it, any time. Start slowly." came the answer up the tube.

When the anchor was at the water's edge we started at half-speed. When the anchor was stowed, we moved to full speed with another belch of black smoke; we had "a bone in our teeth" (a huge bow wave). I eyeballed a course which would vector us on the flagship as she passed by Grande Island (by the way, in the narrowest part of the channel). We headed for the ship astern of the flagship. And we got up there smartly; they were going slowly, six knots. We were doing double that at least, more than that. We pulled up parallel to that ship on our starboard and dropped speed to conform to her six knots. When I got a feeling of that speed and distance and the space we had ahead, I was ready to move. Our position now was on the left side of the column and our instructions were to be on the right side of the column, abeam of the flagship. So what I had to do was thread between the flagship and the next ship in column. They were close at that modest speed.

For some reason I have never figured out, I felt I had to do it, right then. I called for flank speed (that's beyond full speed) and we raced up the channel on the left side of the column. Another pall of black smoke from our reluctant engines. When our bow was even with the flagship's stern I ordered "Hard right!" And we pointed slowly toward the flag ship. That agonizing moment when momentum is greater than the rudder's bite, it just seemed like we were going to crash into her stern. But we couldn't because slowly we were turning at right angles to the flagship and they were going away at six knots. We swung right by, just missing the flagship, but safely ahead of the next ship in line, which was our real danger.

Actually it never occurred to me that it was a real danger because I knew the ship. I knew the ship's turning circle and the time it would take, at any speed. So we roared through the gap and then I called for left full rudder before we were through so we would not overshoot and be too close to the shore on the far side of the column. It was a very narrow channel there. And it took exact timing. We did it. We turned back sharply, heeling way over, and we roared up abeam of the flagship, then slowed to six knots. Abruptly we settled down in the water, lost the bone in our teeth at exactly the right spot. And I had the signalman blink over to the flag, "YMS 353 reporting for duty." HA! We started our sonar search pattern. We were going to be right in the middle of a line of escorts on this trip back to Leyte.

I am appalled at my temerity when I look back on it. I was appalled even then. We could have gone astern of everybody and finally caught up on the far side. But we had done what PT boats and loony destroyer captains do, not plodding wooden tubs like ours. I was so proud of that crew. But we just settled down to our routine; we knew the drill.

Then the signalman called to me: "Signal on the flagship, Skipper." Our call number flags were snapping from the flagship, from their signal mast. There hadn't been any flag signals earlier. Why us? Only us? All the other ships' captains were surely studying those flags, wondering what was up. Then there was another flag, that I did not know. Why was the flagship doing all this in public, for the whole fleet to see?

"What is the fourth signal flag?" I asked the signalman.

"Skipper, I looked it up. It reads, '353, Well done!'"

There were grins all over the bridge. I felt very strange. The flags came down. Everybody had seen them and the word spread through the ship. I went below to the wardroom, big grins meeting me at every turn. Pride showed in those grins; a veteran crew had their due. And we slipped smartly into our escort routine. Ping-ping-ping. Steaming as before. Our minds had turned forward. Where will our next invasion be?

The flags were down now but, for a moment, the message hung in the air: "YMS 353, Well done."

Stone Age Friends: Ulithi Interlude

FISH NETS IN THE ULITHI LAGOON

11. Stone Age Friends: Ulithi Interlude

Your professor Dad starts this chapter with an examination question, an exam question for the captain of the YMS 353, Spring, 1945: "You have just been through some rugged weeks of combat and have returned to Ulithi to prepare for the Okinawa invasion. Below are some options for your next assignment. Select one of the following:

> **A.** Be scrubbed from the Okinawa invasion (mechanical troubles).
> **B.** Remain in Ulithi on routine patrol duty; rest and repair ship for the eventual Japanese homeland invasion.
> **C.** Be assigned as Naval representative to the King of a tropical island.
> **D.** Take time out to study coral reef biology.
> **E.** All of the above."

"Would you believe, 'E,' all of the above?"

"No."

"Unbelievable."

"Yes."

Here is how it happened.

We had left Leyte Gulf in the Philippines and all the horror of the last couple of months; left in February, '45. We were an escort of a convoy, empty convoy, empty ships. Back to Ulithi. On the radio we heard of the battle of Iwo Jima. We missed it. Pleased.

But at Ulithi I quickly went to a meeting and was handed a massive top secret volume. It was the Okinawa assault and invasion plans. Our sweeps were to go in among the first, of course. And I memorized the Okinawa book. Everything that had to do with sweeping, I committed to memory. It was staggering. This book had everything in it. Operation day by day, what the duty of each ship would be. The fuel schedule, water, food. Date, time, latitude, longitude; where everything was to take place.

There was a sweep plan, ship by ship. There was fire support from the cover fleet, again the old battle wagons sunk at Pearl Harbor. And the CAP (Combat Air Patrol) was to be furnished by carrier fleets nearby. Unbelievable, just incredible, hundreds of pages. Maps, tables, all the details; more complete than previous plans.

I have never been under such pressure. This, and other such operation plans as I'd had before, gave me a spectacular motivation for learning. Almost every detail was in my mind. I didn't need to look at the book, although I had it on the bridge for reference when we were in battle. This was an extremely intense situation. Nowhere, at no time had I been such a good student!

While I was studying, everyone else worked on the ship. Got a new screw in dry dock. Just before the departure for Okinawa, near the end of March we had trials to test the engines and the new screw. A visiting engineering officer from the tender came along and so we went to full speed; it sounded sweet. And all of a sudden there was a thunk, a thunk that just shook the ship. I knew we'd hit something, with the stem, the bow. We looked all over, didn't see anything. And then, rising in our wake, a huge log. Actually it was what was called a camel, a squared-off, enormous log with chains on it; they were suspended over the side of a tender for ships coming alongside; it was a huge fender. It had gotten water-logged; no one had seen it, and we hit it, not with the bow, but with the screw again. Back to the tender! The engineering officer was sick. The shaft was bent and we only had power on one screw, about eight knots.

We were scrubbed from the Okinawa invasion just when we were poised to go, psyched up. It was only a couple of days away. Incredible psychological rebound from that. We were disappointed, truly disappointed. We were ready to go. On the other hand, we were extraordinarily relieved! I don't know which took precedence in our feelings. I took the book over to another bewildered YMS Captain and they went in our place.

We lost a lot of sweeps at Okinawa, several to a typhoon. Many guys on the ship were very fatalistic about that, and they said, "We would have gotten it too." But I had this, by now, incredible belief in the ship and the crew and myself. Somehow I had a feeling that we would have made it. But we didn't go. We were crippled, anchored in a vast lagoon a thousand miles from anywhere. And we were assigned to the atoll defense patrols; even on one screw the YMS could do a patrol.

Ulithi was now so far back from combat that we didn't have a lot of air alerts, just an occasional one. Once we actually had a plane show up and try to make a *kamikaze* landing. Otherwise it was a classic boredom.

I found the log books for this time. For instance:

> **0000.** Wind east, Force, one. Barometer: 29.73. Wet bulb temperature 88. Dry bulb temperature 86, broken clouds. Anchored as before, nearby to Azor Island, Ulithi Atoll. Examined and found to be correct, Richard B. Bovbjerg, Lieutenant, JG, USNR, Commanding.

That was all that happened that day, "Anchored as before, nearby to Azor Island, Ulithi Atoll." The essence, the absolute essence of our duty in the Pacific.

We came in, fueled up, lay to in the harbor, had a little liberty, drank a few beers over on the shore, then back out on patrol. Some other things did happen. We were raided of personnel. I lost my executive officer; we were very close. He was so very good. On leaving he said, "You know, skipper, I have to tell you this. I worked for lots of guys, but I never had one in whom I had such confidence. I never doubted for a minute you knew what you were doing." That touched me deeply. I could only shake his hand, so sorry to lose him. We were buddies. To lose a buddy is to lose part of yourself.

A new shaft was on order from Mechanicsburg, Pennsylvania, by slow boat. There we sat. Out—in, out—in. Then one day, in June, a strange order came for us. I had a meeting with the island commodore over on Azor Island. And he said words to this effect:

> We were to go down to southern Ulithi, off the island of Fassarai [he said Fassarai, but there the folks pronounced it Fackarai]. Anchor, and police the island. We would be support for the native population, and King Ueg is down there and I was to meet him. We would radio anytime there was an emergency on the island.

I should make such inspections as necessary. The one important job, aside from being an emergency radio for the island and King Ueg, was to prohibit visits by anyone, including shooting trespassers after a warning.

I was now in the diplomatic service. We were told, "You don't need to do anything. Just sit there." The sub-chaser that had been down there before us could not wait to get away. They tore out. This was their worst duty ever, doing nothing, not even on patrol.

It took us a couple of hours to get down there. It seemed so strange to sail south, past the entrance to the lagoon, to the atoll. Here were the islands all around that we'd seen from the sea. The masts of a hundred ships or more, up there in the northern anchorage, dipped below the horizon. We were absolutely alone with that island, Fassarai.

We crept into a very safe but very close anchorage, between a couple of neat coral patches. It seemed like we were right on that excruciatingly bright beach, but surely we were a few hundred yards off. These were placid waters. We knew the outside of the island, with the pounding surf and the boom. But here was brilliantly clear, quiet water. We could see every coral patch, every school of fish. So we dropped the hook. The engines sputtered off. And then there was just the gentle hum of the generator engine, the Budda. We set a minimal watch. Everyone hit the sack. The bottom dropped out for us. We were at the end of the world. My God, how long would we be here? Weeks? Actually, it turned out to be a few one week assignments interspersed with replenishments and patrols.

It looked like a tropical prison to the crew. We were allowed to go back up north after a week for mail and provisioning, but as you might guess, I was terribly excited about this and it did somewhat rub off. We had lots of diving parties and the necklace and ring manufacturing picked up. We heard guitars on deck and there was a lot of poker playing, a lot of sleeping and relaxing. This sure as hell beat war! That was the alternative.

The orders said to pay a visit to King Ueg. I could not wait! What an unbelievable opportunity! No one else in the world could do what I was doing right now—in the world. I rowed in by myself. I was in tropical khaki shorts and shirt and sidearms, gun belt, this was an official visit. I jumped into the water with my shoes on and hauled the boat up on the sand. I could see the tide was out. The ship seemed small, seemed so far out on that vast lagoon. There was one pennant flying, a black and white striped flag that flies the minute the skipper leaves the ship. I could barely see it flying there. "Commanding Officer not aboard."

A fairy tale tropical paradise was here. And I was going to pay my respects to a King of this island.

It was a narrow island. The beach was spectacular and very broad. Through the palm grove I could see across the island and hear the reef. It couldn't have been more than a hundred yards or so. I had landed near a huge house. The roof was steeply sloped, but it also pointed forward, leaning toward the lagoon in a peaked prow, a big peak reaching out.

A young man met me with a huge grin. I was expected, of course; they saw us arrive, they knew the drill, they'd had ships out there before. He spoke good pidgin English. His name was Edwourd Mozoleh (he could scrawl it out on paper) and he said to call him, "Eddie." He was just a neat guy, I guess sixteen years old or so. He took me to what turned out to be the All Men's House, where strangers were greeted.

King Ueg, a paralytic, sat in a wooden cart that had two bike wheels. The Japanese, who had been here since the World War I mandate, had made him this cart. He smiled and nodded his head. My God, the dignity that man exuded sitting cross-legged on that cart. He clapped his hands and two little boys went dashing off and came back in seconds with huge green coconuts which he chopped with his machete and took the tops off. Since these were green nuts, the white meat of the nut had not formed and it was filled with milk. It was sweet, cool milk; it was clear and it was delicious. The King nodded his head to me, held up the drinking nut and sipped this stuff; so did I and this was a very formal sort of toast, and we bowed and Eddie translated. The king did not try to say a word in English; he smiled some understanding.

I found out immediately that a Navy doctor had spent months working with these people, living on the island, had a sleeping tent there and a hospital tent; he taught English to the kids. He'd been gone for months by the time I came but the kids did know some English.

It turned out that the Navy took a very protective view and I applauded it the minute I realized it. All the native people from the several islands of this atoll were down here towards the southern end. They were on two islands, of which Fassarai was one. Here they were not to be touched by Western culture. Out of sight of that war machine and you might say out of reach of the sailors. There was to be no destruction of the native culture. This was a really very enlightened military position.

The doctor was sent there for their help and he was the exception. No way, however, he could avoid having an impact on that culture. He had a small diesel generator for lights and a radio. He left several teenagers with ambitions outside of their culture. They were altered by that visit. This was a man of tremendous good faith and good will and hard work. He did not want to destroy the culture but his being there in a sense started it. Eddie did routine upkeep of a diesel generator!

We finished with the formalities and I was introduced around to some of the elders. I had landed at the right place, at the All Men's House. It was the formal site for all entry, for all visitors, the political house of the men-elders, no women allowed. There were canoes drawn up in the shade of the trees, off the beach. Logs, coconut log rollers helped in dragging 24 foot long, gorgeously graceful boats up into the shade.

Eddie took me on a tour. The island was actually a village. There was no thick jungle here. It was open, almost entirely palm trees, a few bushes. There were thatched houses, but not in rows, although down the center of the island was a rock-bordered path I'm told was left by missionaries. That was a thoroughfare but the homes didn't border it in any orderly way at all. I met my first villagers (there were about 200 people on the island). They grinned widely and greeted me. No bowing, no servility but no hostility. The men wore only a breach clout, wrapped around the waist, tucked under the crotch and in the back. Women only wore a thigh-length wrapped skirt and were bare-breasted. I could not watch them, and turned my face. Whenever we got close to a group of women I looked away. It was an intuitive thing; I was a creature of my own culture.

It was breezy. It was lovely. On the windward, ocean side there was a dense belt of pandanas trees. These are not very tall trees with long palm-like leaves and stilt-propped roots. There were a few bushy trees along there. On the lagoon side it was open, palm trees and a few bushes. Here's where they all lived. This was the village and I was seeing this for the first time and I asked myself, "How many other people in the world have ever been to this place?" (Turned out to be more than I thought. There were earlier visitors that came to see the doctor and bare breasts). North of the village it was more forested. There were palms but there were some bread fruit and *Pisonia,* big hardwood trees, and some cultivated taro.

It was hot and it was quiet. There were only "the natural sounds." Voices stood out, one voice at a time. The kids ran and stared. I could hear insects buzzing. After about twenty minutes I started smiling back at the women; there was absolutely no way I could go on avoiding looking at women every time I met them.

I noticed there were so few young people. There were little kids, there were a lot of urchins but there weren't any teenage kids, young adults, only three young men. I asked about this and Eddie told me the Japanese had kidnapped all the young people and taken them to Yap, several hundred miles to the west. Here they would go to school and become part of the Japanese "Co-Prosperity Sphere" plans for the Pacific people. Apparently this boat load of young people were told they were just going to go out for a ride, going to another island in Ulithi, but instead they went out through the channel in the reef. When they were going around Falalap Island, two of the young men, including Eddie, jumped over the side and swam ashore. Hatti, who was here, got to Yap where he stole a boat and sailed back, those hundreds of miles. One of the other boys got a bullet in the bottom. The king, in later conversations, called this kidnapping the castration of his village and that they would never recover.

As we walked, Eddie stopped to drink and I saw then what all these things were I'd seen on the coconut tree trunks. Hard coconut shells. They cut them in half and tied them around the tree with twine. Every time it rained, the water ran down the trunk and filled up these little cups. You'd just walk up to the tree, tip it into your mouth and put it back: a drink of water. I didn't taste any; I had a canteen. They had some shallow wells; I saw these and they were fresh water, a bit brackish. I now know, did not know then, that fresh water soaked down through the porous coral soil and made a lens on top of the salt water which was deeper and continuous with the sea water.

I asked Eddie about sanitation. I saw women bathing in the lagoon. No problem, they couldn't be more naked than they were walking around the path. They didn't remove their skirt. They just waded out, and bathed and reached up under their skirt. All bathing was done in the lagoon in the shallow water in this way. I saw no sign of soap. There were no latrines and I didn't smell anything; Eddie told me that the outer sea reef was the toilet and the lagoon was the bath. This is so damn sensible because the tide flushing in day after day after day, pounding the water in through the tide pools, cleared the reef clean.

Almost every day the kids would wait for me. I went ashore every day. And a dozen or two would greet me. I just loved them. This was, in fact, their outdoor school. They learned and I learned. About half the day the kids helped their parents in trivial tasks, which was learning, and the other half of the day they were taught by the older kids, in packs. The older kids were the teachers of the younger kids. They learned the names of every critter and every plant on the island and the reef. They learned 1) where it lived; 2) if could you eat it or use it; 3) if it would hurt you. Little "Theek" was a toddler two or three years old maybe. He walked quite well and he toddled on that reef bare-footed. Some seven-year old would have hold of his hand so he wouldn't get knocked over. And they would teach him. I remember watching an older kid pushing Theek's finger onto a spiny sea urchin. The kid got the name and he got the pain, and he got tears. He knew from then on what a spiny sea urchin was.

They all knew I was interested in shells and such. This was a biologist's paradise! This was pure Pacific coral reef. I thanked them endlessly for all the gifts and I pocketed all the booty. Theek knew this, the two year old, and he would bring me every pebble he found, whatever it was, broken piece of shell, broken piece of coral. And I had one pocket with a hole in it; Theek's pocket. And as much as he gave me I lost through the hole in my pocket. His grin was my booty.

These gangs of kids just roamed around learning from each other. They raced around and horseplayed and hollered. And if they ran into groups of women working they were called on to help. They stopped and did this. The older kids spent more time at work, the true apprenticeship to adulthood.

There were three girls just into maturity; I don't know how old. Thirteen, fourteen. Maria, Elvira, and Pietra. Cute, bud-breasted, big eyes, grins; rounded adolescent girls. They were trained by the Navy doctor to run the sick bay and they had sick call every morning. They were not among those who were kidnapped by the Japanese; I guess they considered them too young. They had matured a lot in a couple of years.

Where did they get these names? Elvira, Maria, Pietra? For that matter one of the young men was named Jesus. These names were from Spanish Catholic missionaries, beheaded by the Japanese years before. And Ulithians carried over some Spanish words; they liked some of those Spanish names. There was no trace of Japanese culture that I could see. They had rejected all things Japanese. These were Japanese World War I mandated islands and the Japanese used them for the copra trade, then used them as weather and radio stations. And they brought some cloth for skirts and breech clouts, and of course, the King's cart. Eddie gave me a beat up little Japanese biology book.

I was avid to fulfill my duty; inspection every day. I liked mornings; quiet, peaceful. I made the rounds, looking for those footprints in the sand. There were none of course. I couldn't get enough of the place. Once, Mozoleh (no Spanish name that!) said, "I wish you wouldn't sit under the frangipani tree."

"Why not?"

"Well, it will disturb the spirits."

"What?"

"The spirits live here. They shouldn't be disturbed." I loved the place. The aroma was wild; and you could sit on a carpet of white petals, but I did not sit there again.

I never really got it all straight. Spirits are apparently departed souls that leave the body at death. The essence of life is the spirit. The bodies were put in a rather crude cemetery. But the important thing was the spirits were related to you, and having them around was good because you could go talk to them and get guidance, and people apparently did. I really couldn't get at this, not knowing their language.

Then there were supernatural entities. Demons did you dirt, and they had to be placated. And there were gods in control of all sorts of aspects of their life. And they were to be respected. I found no notion of heaven and hell, no sense of sin and punishment. And I asked Eddie why folks seemed to be so well behaved, and he said, "Because that's the way we are." He couldn't comprehend the question. There was no other way.

On a couple of occasions Eddie acted as interpreter and I had some talks with the King (with the always present green coconuts for a cool drink). He seemed to have a very clear idea of their place in the world, that they were on one island in a vast area of islands. What was strange were all these ships and planes and people. He knew Yap, to the west, but he did not seem to know the Micronesian Islands to the east, over toward Truk. His people dimly recalled the missionaries. They certainly knew the Japanese. And now they knew the Americans. There had been only a few Japanese here, and, and they left when our planes started buzzing and strafing everybody. They went over to Yap. We had been scheduled to take Yap; it was a gorgeous high island with barrier reefs. But Ulithi was casualty free for our forces, a gift. And it was in some ways a better base, as I have described.

King Ueg told me how the island was run. This was fascinating. He stressed how important the elders were. They handled all matters of importance, with greeting visitors at the top of the list, or so it appeared. This was of course their contact with the outside. There was apparently a little inter-island trade and marriage, and that was probably good for the gene pool. And of course the elders acted as judges in any problem, apparently this usually meant divorce. I didn't see any other problems. But apparently divorce was not difficult. Of course the men decided all ceremonies and things having to do with religion and magic. Magic seemed to come in here some where. Men were the providers of protein from the sea. They were the builders; canoes, homes, tools. Above all they were sailors.

Well then, what do the women do? Well the King said "almost nothing." But they are in a sense the head of the families. This society is matrilineal, things are reckoned by your mother's, not your father's ancestry. The women decided on the marriages, I gathered. After all they're the only ones who positively knew the real lineages. They grew the food, prepared the food, cared for the sick, manufactured mats, cloth, adornments, thatch for the roofs, baskets. In other words, the women ran the island, but not beyond. That was for men. There seemed to be a definite male social dominance. And the men thought they ran everything. But there was no outward public show of female subservience to males, unlike many civilized societies.

I asked the King, "How do you treat those who disobey?" He did not understand the word obey. In the King's memory there had never been an assault on Ulithi. Never any punishment. He remembers one woman who was giving the village a great deal of trouble. She was after all the men; insatiable. So they put her ashore on a distant island with no canoe. She would of course survive; she knew the islands and all the tools and skills of survival. But she would no longer be a bother. Now, that was not so much punishment as the removal of a bother, I guess. I asked the King why the people always did the right thing. Again, this gave him trouble, this concept of right and wrong, but he said, "We do everything the way we must do it."

"How about stealing?"

This he could not understand either. He could not understand about people being hurt. It's not done that's all. No sin, no punishment, no applause. Just not done. Taboo is a strong word. Polynesians claim the word taboo, and tattoo, but both were important here.

I saw no child punished. Infants were picked up by the nearest adult male or female, or by another child. I saw no fights, heard no harsh words, even in the hormonal rush of pre-teens, which were there. Lots of tearing around, laughter, teasing. My guess is that teasing was aggressive, but funny. I think some teasing was in a dirty language, somehow, the giggles and looks. I didn't know the language, a pity.

Stop here! I need to plant a caveat. Do realize that this is not an anthropological study. My contact was so limited; never saw the inside of a home, never even spoke to a woman, did not see a birth, death, or marriage. These are purely vivid memories of what I saw and heard in that incredible time spent with these incredible friends.

Eddie Mozoleh was my companion every time I came ashore, I don't know how he got assigned to me by the elders. I did not ask. He was full of stuff. He grinned; he'd lecture on anything.

Then there was the Woman's House (Eddie answered my questions). Nothing like the grand men's house, just a little thatched house. No men could go near, best not even to look. All menstrual periods were gone through here, childbirth, anything else female. Women cared for women. The men were absolutely excluded. I was told that in the first year of menstruation, the girls lived in a nearby hut and simply lived a year of life adjacent to the Women's House. I expect they saw everything. They saw all the menstrual periods, tough and easy, and they saw childbirths. They were out of circulation for a year. This was a womanhood internship. They were then ready to have babies and a husband.

Eddie told me that the brother was not allowed to have any contact with his grown sister, including looking, not till after she married. And you can see that the incest taboo on a small island would be so vital. But then I don't know of any society that doesn't have it; maybe not this strict.

Little kids were very free; they ran naked. A lot of horse play. I had this feeling of sexy teasing among even little kids. I gathered that sex just started when it started but apparently it was very private; I never saw it. Eddie was not keen to talk a lot about it but I could sense it.

I had an interesting experience; I spent a couple of nights in the doctor's tent. It had nice screens and it was very neat and had a light. I slept in my skivvies, the white boxer shorts. And all around this tent was solid open screen. I couldn't count the eyes. They didn't know their eyes shined like that from the light. And there were giggles. They seemed desperately curious about how this guy was put together. At night when I lay there not sleeping the ground cover of leaves and branches just rustled. And I asked Eddie what this rustling was. And he said, "rats and land crabs." I saw the holes and occasionally saw a land crab, but no rats. He said that they made a lot of noise at night. And maybe the kids were out there too, making rattling noises. I never had any advances made to me. None of the girls or women gave me another look. I was on another plane; I might have been an alien species.

The doctor, who was so close to them (and they did love him) saw this same "noble savage." But that notion of noble savage has certainly been proved a fallacy, as a generalization. There are so many places, undeveloped nations, where people would steal and hurt. The doctor said that the gentle Ulithi manner was from Spanish Catholicism, Christ's teachings. But I could not go along with that. There just was no heaven and hell notion here that I could gather. There were some missionary folktales, but barely remembered.

Their altruism was the outstanding trait. And this of course describes Christ's teaching. And the doctor saw this. Everyone wanted to give me something; they would come from other islands with handfuls of shells (the word was out on me!). And big smiles came with this gift. It was just done. There was no reward; it was a social reflex.

My own view on this is: on large land masses, like New Guinea, the clan of one valley would have enemies in another valley—territoriality. Not news in the mammalian world. Fight, kill, steal, crap on the doorstep. Really be mean, even if only in sham fights. But on an island, a tiny island like this, this attitude would have been lethal and the society would not have survived. Love is an adaptive trait in the social sense. I don't mean this as social Darwinism; it is just obvious and common sense. I do think that it's useful to use Darwinian terms.

One of the big decisions made by the elders was, "When shall we go fishing?" I do not know what that decision involved. Something mystical? Then all hands, all men, and boys that were strong enough, hauled out these huge seines. I don't know, more than a hundred yards. And they carried them out by canoe and made a huge circle as far as it would go, make a big loop; it was secured at one end to a palm tree and pulled on the other. This enclosed any fish that might be in the bight of the net. They pulled them in and took out the fish they wanted, threw back the ones they didn't. Then they would string them up on a pole and two guys could carry the food to the feast site. They were doled out apparently by family. I pondered: Now, how do they regulate this commune? I do not know; there just has to be some rank in there somehow. Age, I don't know.

Women cooked. They had glowing hot coral, heated by burning coconut leaves and husks, maybe all afternoon. Then they put in fish, grated coconut, taro and coconut milk and wrapped this in green leaves. Everything was done on the ground with woven mats for a "kitchen" table. I obviously never tasted any of this. They would always offer me fish, and I just had to say no. Every time they had them they stopped and asked, "How many did I need?" Food did not seem to be on schedule and it did not seem to be a problem. The kids were just fed ad libidum, from what I saw.

Never tried to pick up the language. I remember one word, the word for the lizard. It was onomatopoetic. They were called "put-a-put, put-a-put."

There was an intense interest in English, and American things, mostly by the kids. They just soaked it up. And I think this doomed the society. They wanted stories; stories in pidgin English, with gestures. So important, the gestures. And mime would lead to extravagant gestures, buffoonery and laughter. When I told stories I would use a minimum of language and a maximum of gestures, and it was fun. I just loved those kids. And they were so shy and happy and they laughed, so naturally. We chattered.

Eddie wanted to come out to the ship for a visit. And I debated, but agreed. I don't believe that he had been invited to the other ships; he was all eyes, mind clicking. This was a smart kid. And here was civilization on this giant canoe for guns. He saw the guns and he knew what they were for. He used to bark machine gun noises and make stabs at his chest. His uncle was shot by machine gun when the American planes came.

The crew invited Eddie for lunch. Later I showed him a *Life* magazine. We sat down on chairs and put the magazine on the table. What an incredible thing, a chair and a table. But not as incredible as the first thing I turned to in the *Life* magazine. I remember that magazine for only one thing: the first page, inside the cover. It was a Pullman Car ad. There was a train, and there was the inside of the Pullman Car. I tried to explain to him (my God, three thousand miles of track); beds, chairs, tables. And they were running faster than any ship he had ever seen. And they did this day and night. I was sorry later that I furthered the education of Eddie.

I had asked him to bring some coconut husk to show the crew how to make rope. He got a big gang to watch this. The rough fibers are separated into handfuls. These were usually soaked ahead of time in water, like linen, like flax. And then a few gathered fibers are rolled quickly on the thigh. When released they recoil back up the thigh; they twist; the fibers turn into twine. Then another bunch is twisted into the ends of the twine to make a ball of twine. And twine is twisted into a cord; and cord is woven into a rope. The bosun was excited. Rope was after all his business.

"Can I try that?"

"Sure, Boats."

He had on shorts, so he was ready to go. He made the first fast roll down his thigh and bellowed with pain; my God! Each hair on that hairy man's leg was trapped and ripped out by the roots. And Mozoleh grinned. His thigh was hair-less!

I made a policy of rotating trips ashore, field trips for about six men at a time. I thought this was a chance of a lifetime and they agreed. They were awestruck. Not just by the bobbing anatomy (they were told not to stare), but I think far more by the impact of such a culture and society. They saw that this was not a show put on by a bunch of natives. I think that is striking. This is all of it; this isn't just one town, but a whole society. The men were, I think, deeply moved by this, and they cannot have forgotten that to this day, as I have not.

I talked with the three girls who were in the sick bay. This was quite a place, with shelves and cupboards and medicines, tables, an examining table. And the three girls had morning sick call, and mothers, pregnant women, and kids got in line for a vitamin pill. Then they were asked if they had any complaints. Navy sick call. The three girls were the nurses of the clinic, with no doctor, but the doctor had trained them. He was sent here to clean up disease and set up the medical tent. The incidence of yaws went to zero. Penicillin, new stuff. Everything else cleaned up, including suppurating wounds, especially coral scratches. Little six year old Plasenta ran on his horribly scarred leg; he had been a cripple of yaws, could not walk. But the doctor was moved by the Navy to more urgent duties. I asked the girls about each medicine; symptom, treatment, prognosis, and they did very well. They had shelves and cartons of medicine. Of course they understood nothing; they just memorized. How long could they keep this up? Could they keep it up very long at all? I don't think so. I wonder what became of the girls and their "hospital."

These folks had very little in the way of art. I looked all over; I was very curious. They did adorn themselves and their homes with flowers. Some of the homes would have quite ornate bouquets, more like wreaths. And they carried flowers over the ears; and necklaces of flowers. There were so many around. I saw no carving, no sculpture, no drawing. They tattooed; the men were actually sometimes extravagantly tattooed, especially the older men. Most of the body; torso, arms. The women's tattoo seemed to be more private. I got glimpses of insides of thighs and arms, but not on the torso.

They're good looking. They weren't tall. I was taller, at 5'9", than most of the men. There were no fatties, there were no scrawnies. They were muscular and smooth. They ate well, but they worked hard. They had such big smiles, and this composure and dignity which struck me. Their glance, the look of the eyes was level. Innate dignity and composure.

I never had a feeling of superiority over them. Here I came with all this western culture, this baggage, this education from one of the finest universities in the world. This also struck some of the crew. They didn't talk about them as "gooks." These were people with names; Eddie was a friend.

These folks told stories. They had no written language but I think they told a lot of stories. I gathered that the men's houses and women's houses were storytelling places. I so wish I had known the language.

I'm told they had dances. I have since read descriptions and seen pictures of them. I wish I had seen them then. Polynesian dancing is so spectacular and these are probably related. They had no singing that I heard. But the doctor apparently had taught them a few tunes. And one hit me like a ton of bricks when I heard it. I heard little kids' voices, "Marzy doats and dozy doats and little lamzy divey;" it just cracked me up. About two years earlier this had hit the United States. They had no idea what a mare was, a doe, a lamb, oats or ivy. But they loved the lilt. And they just sang it and sang it and sang it.

Certain men made canoes and this was art as well as craft in making this very important tool. I made some careful measurements and drew a scale drawing of it, which I still have. They had an adze with a shell blade at the end of the wooden shaft. They sharpened that. That crude blade must have just been hideous to use. The Navy had intervened however: had gotten them plane blades or something like that. Now they could really whack out chips. They cut planks and then fitted the planks together. The hull was V-shaped in cross section, very narrow, and very deep; it had marvelously slick streamlining to it. So graceful. There was a lift at the stem and stern with a wide fork for a steering paddle. The planks were put together without hardware. They drilled holes and sewed the boards together with cord made of coconut husk (or was it shark or turtle leather?). And the cracks in the holes and the cord were "tarred" with tree sap from the breadfruit tree which was chewed until it got like bubble gum. They stuffed it in all of the cracks. It stayed soft apparently and sort of joined these boards together. It could obviously be redone at any time. They made lateen rigged sails and these were of woven pandanas leaves. Easily repaired. The rigging was from coconut fiber rope, and the anchor was a hunk of coral with a hole in it. There was one large outrigger. And the boats just seemed to sit high on the water, as though resting on glass suspended over the coral bottom, really gorgeous. Ulithians were spectacular sailors.

They however were amazed at my little boat, the little wherry. It was fitted by the first skipper of the 353 as a little cat-rigged dinghy. And they just thought this was spectacular, especially because it could turn on a dime with a flip of the tiller; it didn't have this great big oar back there which they had to lean on to turn. My tiller just went "whoop" and we would whip about; the wind would go from one side to the other so fast that they just screamed with delight at this thing.

I was however saddened by all that I saw because it was clear that there couldn't be a survival of this culture. And I hear indeed that they have not survived as a culture. Now, I am told, all the kids want to be electricians in Truk. They know about Truk now. High island, hundreds and hundred of miles to the east. They want to have a motor bike and drink beer. Why eat fish and coconut so you can sleep until morning? Then gather more coconut and fish?

A fellow from UCLA, Bill Lessa did a Ph.D. dissertation right after the war. That would be very shortly after I left those islands. I read his stuff with interest. He saw what little I saw; but he of course, saw so much more, learned the language. He has a very nice little book out, which is really well done, and written for the laity; it's called, *Ulithi: A Micronesian Design for Living*, from Stanford University "Case Studies in Cultural Anthropology." I certainly recommend it. It was such fun for me to read that and see his analyses of these things that I saw. That has to be part of what I tell you here.

Then a few years later, my friend at Stanford, Don Abbott, and Marston Bates, a well known ecologist from Michigan, came to see me at St. Louis. They were going to go to Ifaluk, the next island east of Ulithi. They did an extended study and wrote a book about it. These were very fine biologists and studied the ecology of the whole island—but, importantly recognized as well the biological position of the human population. Those people on Ifaluk in the 50's were still primitive. There had been no Navy there in the war, but there were indeed some of the things that the Navy subsequently left them, but not the contact that people had at Ulithi. But my colleagues in anthropology tell me that this way of life is gone in Micronesia. A recent book by Catherine Lutz on Ifaluk emotions suggests a measure of the old culture remains. Fascinating that the title of her book is: *Unnatural Emotions*.

Had a very disturbing experience one afternoon. It was such a story book afternoon, the sky so incredibly blue and the cumulus clouds so incredibly fluffy. The reef boomed and I could hear the children laughing, and the insects buzzing in the flowers everywhere. I just relished this atmosphere. But suddenly everybody was cheering and hollering. I looked and a Navy torpedo plane buzzed us. It came low over the trees; it roared through. They were spraying. For heaven's sake, they were spraying a big gray cloud. It stunk. We all coughed and ran. The wind carried it away finally, but we were stunned by this; we'd been attacked. I rushed back to the ship and radioed up to the north, up to the island command.

And they explained it, they said, "Oh, yeah. We're sending this stuff down now. We got some brand new stuff, it's going to kill the insects and stop the diseases."

"What is that stuff?," I asked them.

"I don't know for sure," but, he said, "it's a miracle spray and it's something like DT, or something like that." he said.

My orders to leave did come. They came when the new shaft arrived from Mechanicsburg, Pennsylvania. We were granted availability with a tender, the floating dry dock, and all the routine up north. Time to ready ourselves for the Japanese homeland invasion.

I knew that I had to formally say good-bye and it was painful. Presents were clearly in order, but what in the world to give these people? The crew knew. In my inspections I had seen in a galley cupboard a huge, heavy, silver-plated serving tray. This was part of the officer's pantry. We had a bunch of this stuff that we never used. I marked down on the inventory, "Damaged and jettisoned." What a present for the king. And the galley gang polished and they wiped and they scrubbed and it just shone. Oh! And on this we put folded cloth. In a compartment under the deck on the bridge I remembered that there were bolts of colored cloth for repairing signal flags. What an archaic idea! A ship this size has bolts of cloth? It doesn't have a sewing machine, doesn't have anybody who knows how to sew; we never used the stuff. Heavy duty cotton, colorfast, royal blue, yellow, red, white and black. I cut the cloth into six foot lengths. Breech clouts for the men and skirts for the women and a black one for the King. He alone would have this black cloth.

Then I thought, "Hey, I have some cigars." In my locker was a box of Cuban cigars from Havana. They were a year old. Still wrapped in cellophane. For the King and the elders. What a way to say goodbye to all these old guys. I dressed up in a cap and everything; the King sat in his cart in the All Men's House and I squatted on my haunches (I got good at that). We made speeches, undying memories and all that Bah humbug. There was formal toasting with drinking nuts. I was choking up.

Then the gifts. I have yet an adze handle that they gave me. They gave me a hair comb which I gave to our Anthropology Department. The king's eyes sparkled when I gave him the box of cigars, completely wrapped. He pulled apart the cellophane, opened the wooden box, and there was an iridescent green stack of cigars, covered with mildew. He offered them all around, and of course me first. We lit up. It smelled like a good cigar and it smoked like a good cigar and it tasted like one. Everybody was wreathed in blue smoke and big smiles; there were no ill effects. It was a suitable cap to a very deeply felt goodbye.

I went to my boat on the beach; launched it with lots of help. And the elders were standing there, just standing, each on a pedestal of dignity. I rowed out. All the kids raced down the beach, that stunning bright beach. They were waving and shouting. And in the bushes, partially hidden, I could see all the women waving and smiling. They were all saying goodbye, differently. And the ship got closer and the beach receded until it was finally just a line of graceful palm trees; I could see some dark bodies. And then I just couldn't see anything, not with my eyes.

Looking back, what do I think our greatest presents were? Not the little ones I gave. The greatest presents, given by well-meaning people, were our contribution to the erosion of their culture and the marvelous miracle spray which undoubtedly was one of the early uses of DDT.

"For Those in Peril on the Sea": Typhoon

ULITHI ATOLL

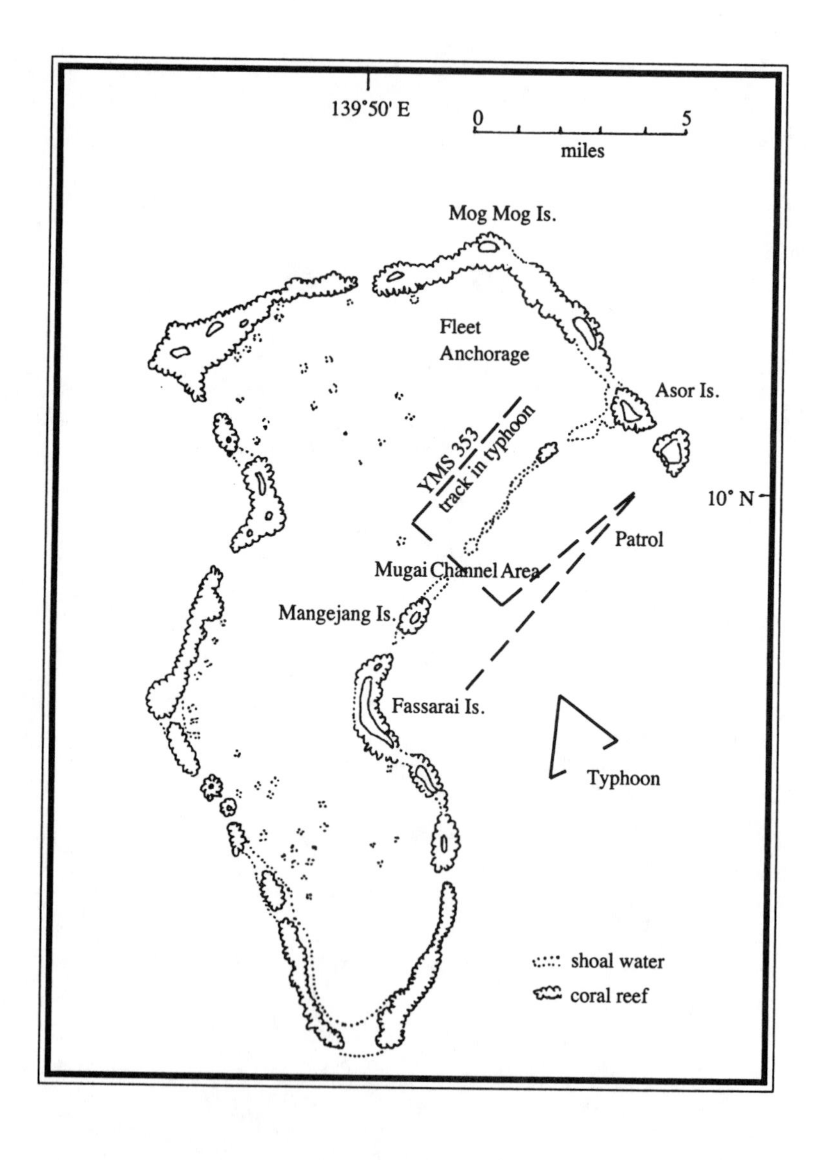

12. "For Those in Peril on the Sea": Typhoon

There are two things that bring uncontrollable lumps to my throat. First of these is "Taps;" second is "The Navy Hymn." I'm entitling this chapter for the latter, "For Those in Peril on the Sea." This is our story of one typhoon.

It was in the summer of '45 which was spent in a rear area, Ulithi Atoll. We were a minesweeper for that huge, floating Naval base, but we were actually at sea much of the time. We did anti-submarine patrol. We guarded the passage in the reef, Mugai Channel. It was in the middle of the atoll on the east side, and all the traffic came in and out of this one opening. It was a buoyed channel, about a quarter mile in width, a booming reef on each side. Always you could see the white water and then the shooting surf that hit this reef. Mangejang Island was one mile to the south which made a very good land-fall. (Check map)

We patrolled across that entrance in Mugai Channel, ten miles north and ten miles back south. Then we'd go five miles north and six miles south and eight miles north . . . back and forth. We were about two miles from the reef more or less; sometimes closer or out farther, constantly pinging for submarines. We got all kinds of echoes. We got echoes on the reef. We got echoes on ships going by—never a sub. The courses were intended to be non-predictable, but still cover the area. Occasionally we patrolled as close to the reef as we could get. That was impressive; shark teeth to a ship.

We were now the rear area. We did this dull routine for months. We'd go out for a few days and then we'd be relieved and return for supplies and fuel and water. We really knew those islands, that reef, those waters. No great danger. We did a dawn and a sunset alert. We were in the habit of that from days in combat. We had a rare plane attack, a few alerts. It's a long distance to Japan, which was absolutely straight north of Ulithi. One night a carrier was hit by a *kamikaze*. They were showing a movie on the flight deck. Can you imagine that? But we felt so secure. The cruiser *Indianapolis* was sunk by a sub just north of us. We didn't know it; nobody did.

The sea can also be the enemy. The typhoon. We had been through one here in Ulithi on the way out. We had seen the results of typhoons over in the Philippines. Some had passed not too far away; we were in the outskirts of typhoons. We were in the middle of this one. And we got the big one. One hundred and forty mile-an-hour winds. That's the top of the Beaufort Scale. It doesn't go any higher.

This caught us out on patrol. There were hundreds of ships in the lagoon safely at anchor. We could see them; at night we could spot them on radar. There was a two day buildup and a two day wind down. We had five days of over full gale force winds.

First warning was the "glass," you know the old sailor's term for the tall mercury barometer; the mercury level fell. Then the wind picked up the waves, even the small waves; the wind was going so fast that it tore them. Spray flew from the top. The whitecaps were decapitated and sent flying over to the next wave and beyond, sheets of froth, spume and spindrift.

Then the rain. It started as a summer rain squall. Big black cloud in the Southeast. But then the whole half of the sky blackened. Then finally it came over us. It was slow; it seemed fast. The barometer fell more. We now had a full gale and a deluge. Patrolling continued but damned uncomfortable.

Finally we got word by radio: "Typhoon. Every ship to take routine measures." No word for us. Barometer fell some more. The waves got steep-sided; they began piling, piling up and becoming very confused. And the trough between was not wide enough to hold them. They became unstable and crashed in all directions. The wind tore at them. The terrible specter for us was the lee shore, the downwind reef, only two miles away. That was wicked coral, and death rather quickly. The rain started to fly parallel to the sea; it was horizontal. And it was mixed with salt water, tasted salty. Sea and wind were still from southerly.

The ship began corkscrewing with the following sea and going into the sea it bucked wickedly. One leg of our patrol we were heading south into the seas and on the other leg to the north we had a following sea. Heading into the huge waves, solid, green water came over the bow, would rage, tear, and tumble in cascades of foaming water the length of the entire ship. The scuppers weren't enough (the holes in bulwarks—the solid railings); the water shot out of those scuppers under pressure. Water was deep on the deck; it was heavy. When we rose, it would all get thrown off. We had only six feet from the main deck down to the water. That's too close!

All the hatches were dogged down, and the ports. All the deck gear had extra lashing. A loose piece of gear would be murder. We lashed down our depth charges with extra care and made sure that they were set on "safe." Terrible thing if they should roll over the side and blow us up. Life lines were strung up all over, always one in reach.

Wind came up over a hundred miles an hour. "All hands put on life jackets." We couldn't stand anywhere, only crawl out on the main deck. The wind blew you down. Or you could clutch a line or climb up a ladder on hands and knees. We secured the galley. Couldn't cook but we had sandwiches and coffee. The cook was busy. Zero visibility; thank God for radar.

One huge ship turned into the channel where we were, a large, bright blip on the radar screen. I didn't know the skipper's intentions or where he was going to turn. But he suddenly did turn and we were right in his path. We were heading on one course and he was heading on the reciprocal and there was no way I could avoid him. He did not signal us. He didn't sound his whistle; he didn't call me on the radio and tell me to get the hell out of there. Nothing. Abruptly this ghost of a cruiser slid by. And she was so serene. Waves were flying from her bow. But that enormous ship, a city block long, was so steady compared to us. We'd go up one wave, down into the next trough. But she went cruising by like a big cigar, straddling the waves. We saw the bow, then the bridge, then the stacks then the stern. We never saw the whole ship at once. That's the kind of zero visibility we had. A close one!

The sonar gear signals finally became too confused to pick up any echoes at all. We really could no longer patrol. There was the blackness of night in the afternoon. We went actually from a good fighting ship with lots of experience and security and confidence, to a mere survivor. But no orders came to leave our station. I figured there were sixty foot waves, probably more than that; they were taller than any part of us. And the wind got worse. It set up an eerie howling because all those masts and stacks, everything had guy wires running to them. This howling was added to the crashing waves and the rain pelting, hammering on the metal stacks and canvas gun covers. It was pandemonium.

We heeled over forty-five degrees in the trough. You realize that at forty-five degrees it would be just as easy to crawl up the bulkhead as the deck! The galley was awash. Someone would come in, slam the hatch behind him and admit a bunch of sea. Crockery was broken. There were books and magazines and playing cards, clothes sloshing in the water. The water would surge from one side and then to the other with each roll. The drain sort of worked. Everywhere a stink. Our bodies were black and blue from just stumbling and bumbling around; and our knuckles were gashed and bloody.

After two days there was no longer any duty to the Navy, just to the ship. And I called ComDesRon on the radio (Commander Destroyer Squadron), "Can't keep station or operate sonar. Request instructions."

The answer came back fast. "Commanding officer use own discretion," which was what I dreaded. But I had made a decision on what to do. And I had selected from three choices. We could head straight into the sea, straight south or east of south. Just head into it and steam about fast enough to keep pace, but get off the lee shore, downwind to that nearby reef. That was all right unless we were to break down. Then the lee shore would claim us. I didn't like that option.

Another, and very seaman-like thing to do, would be circling around the atoll, to the north side, in the lee of the reef. The southeasterly howling winds would still be there, but there would not be a lee shore to pile up on. And the waves would be smashed down by the atoll. Nevertheless if we should founder there, if the engines should go, we would just drift indefinitely until we broached, abeam of the wind in the trough. That would mean capsizing.

My choice was the third option, to make for the atoll passage as usual and anchor inside where we had a good anchorage. Twenty fathoms, good holding, lots of good company. And we would be protected from the monstrous seas, if not the wind.

Nevertheless this was a very serious choice because there were two places where we could die, all hands. At the time the decision was made, we were heading northerly, away from the wind and sea. The first dangerous maneuver: we had to make one last 180 degree turn in order to head back to the channel. And that would put us in the trough. We'd already gone over forty-five degrees; this could mean capsizing. And now the wind was higher, the waves were higher. The second awful maneuver: we had a narrow channel in the reef with a 120 miles an hour wind abeam of us driving us to leeward. The reef would rip us if we misjudged that. The Captain alone could make the decisions—how and when.

I went down to the chart house from the bridge and picked the point of our 180 degree turn. Then I picked a course back to put us two miles off the reef, abeam of the channel. This was two miles off Mangejang Island on the reef, which I had on the radar. When we were two miles off Mangejang my calculation said to turn sharply to starboard to 300 degrees true to hit the center of the channel, figuring in a terrific set to starboard from the wind and seas.

We approached the point for the fateful 180 degree turn, the first awful decision. But before I did it I went up to the flying bridge. I crawled up there, the highest place on the ship, and I searched for the largest and the smallest waves up to windward, coming at us. There's an old saw at sea: every seventh wave is the big one. When we topped a big one, I saw a little wave four or five waves away, upwind. I decided!

I hollered down the tube, "Left full rudder, starboard ahead full." That meant the starboard screw was going ahead full speed pushing us around. The port screw was still at slow. And that was it. I did not back the port screw under that stress—did not dare. There was now nothing to do, I mean nothing. And of course nothing happened. On a ship, even as small a ship as we were you put one screw ahead full, with full rudder, and you keep plowing ahead. Plowing ahead, plowing ahead! It takes so long to turn (I've said this so often—it is critical). The first wave went under us, the second wave went under us. Then we swung a bit. Everyone was up; everyone had a life jacket on. I'd called all the officers on the bridge; we waited. They knew the wheel was hard over. I felt a terrible pang; the men down there had no idea about this decision. Thirty-five guys, their lives were on the line. It was a terrible decision. And only the watch up there on the bridge knew how close we were right this minute, this second, to eternity.

Then the cook staggered out of the galley with a bucket of garbage. He was singing, this hillbilly from Kentucky. The dumbbell threw the garbage and it flew back in his face and he got drenched. He burst out laughing and staggered back in. I heard the hatch of the galley slam shut and I thought, "Oh, my God, the guy doesn't know, in his ignorance, of this moment. He could be dead in less than a minute." It was a terrible moment.

By the fourth wave we were biting fast and we started to roll in the trough. My God, we took a terrible roll. Just awful. And there was bedlam on the ship. I could hear crashes all over. Everything was lashed down but everything broke loose. The men were shouting and cursing. And then, that smallest wave came just when I'd hoped. We were exactly abeam after the worst roll we'd had. We lay over on our side, and stayed there, and then slowly righted as that little wave finally came. And when the little one went under us we started to turn, to bite into the seas. We took another terrifying roll but there was a little corkscrew in this one. And then more of a corkscrew. By the third wave after the little one, we started to pitch and buck. We now headed southerly into the storm. There were shouts and some nervous laughter. Very white faces looked up the ladder to the bridge. They knew by now that we had come close; anywhere on that ship they would know. How horrible it must have been in the engine room, to go over forty-five degrees and hang there like that, trapped below in a potential coffin.

The next decision was made from the chart. It was no less critical. But I had Mangejang on the radar screen and I read the chart, made the calculations again. We were slowly heading toward the island. Then at two miles off the island we turned west to 300 degrees true. Once we started that turn we were in fact driven about by the wind. We snapped around! Now the track should be dead ahead for the channel. And in twelve minutes we should see the reef to both sides, right in the center of the channel. Again, nothing to do. The decision was made. I knew the calculation was right. We'd been in and out of that channel before. Did I figure the wind and waves right? Recall, all of this was in blinding rain and darkness; we could barely see beyond the bow.

The Exec was up on the flying bridge. I stayed down at the radar screen, watching the island's bright blip. Twelve minutes passed; there was nothing to see. Then came the voice from above. He said, "Sk-sk-skipper, skipper. Red can, red c-can to starboard! We're home!" I staggered to the starboard wing of the bridge. There it was, oh my God! It was the textbook solution. There were cheers on the bridge.

Then plunk! Suddenly we dropped into the lagoon, inside the reef, and where there had been sixty foot waves there were ten foot waves. Peanuts! We turned north to the anchorage, but we were safe now. We could anchor. The guys crawled out on deck in the wind. They popped out from every compartment. We rolled so little. It was just so gentle. This was nothing. The men yelled and laughed but the screaming wind tore away their sounds. Only a few knew all that we'd gone through. The bridge gang knew. I knew. There were some very warm handshakes. The Exec came down from the flying bridge—grinning.

We went up to the anchorage slowly. We saw no ships but the radar did. No visibility; the height of the storm. The men crawled up, "stood by" the anchor then got the order to let go the anchor. We faced the gale and were driven back till the anchor held. The engines were left on idle ahead to take the strain from the chain and we had a lookout sitting on the chain so that he would sense through his bottom if the chain were bouncing, which would mean dragging anchor. That's what we did not want.

We went to a watch and watch. That is, half the crew was on watch and half could snooze. The bridge was manned as if we were underway. There was always an Officer of the Deck; constant radar bearings on other ships; distance and bearing of hidden ships, and bearings on the islands, which were fixed. We did not swing to the futile tide, just to the frightful wind; so all the positions of the ships were quite fixed. Every ship was straining at its anchor. The pandemonium of sound and wind was at its height about the time we anchored.

But what blessed relief not to have those huge waves. We only had this short, violent chop. The ship trembled to that terrific straining on the anchor, a force that we could feel. Our whole violent world was out of control and completely intimidating. But what an indescribable relief to have a nap that night. We had done it. Not a wound to anyone; just bruises and scratches, sprains and strains. We had deeply scarred minds however; it was very hard to settle down. We slept that exhausted sleep.

At mid-morning it all stopped, instantly. Utter stillness. Everyone on the ship tore out on deck from whatever compartment, from whatever they were doing, asleep or awake. Brilliant sunshine blasted on all those ships. Absolute silence. We were in the eye.

This inspired awe.

There were cheers and there was a kind of embarrassed laughter. I'm sure there were some prayers. Only a few of us were even aware that the whole thing would return, in mirror image, as the eye went past. I doubted if we would have an hour of this phony quiet. It was a once-in-a-lifetime experience.

We could not wait to turn to for damage control. It was an intuitive thing to start grabbing things and putting them back in shape. There was a cleaning up frenzy, every compartment. Clothes went out to air. The engines got a rest; how they ached. We took visual fixes on the islands. I remember at the time thinking, "I can't describe this." I still can't describe it although it remains incredibly vivid forty-five years later, including that overpowering euphoria which just gripped us. The relief, it was unbearable.

The wall of clouds that formed a circle around us must have been miles high up to a blue sky. The sunshine above was brilliant. I felt like a flea looking up the side of a giant black crock. The clouds were black but crawling with green and brown smudges and streaks. It glowed in the sun, a looming wall, writhing and roiling and twisting. We couldn't actually see the circular movement, which I knew it had, but we could feel it. It was a living thing.

We saw the oncoming wall of horror as the typhoon eye slowly moved across us. It was a surrealistic scene, not in human comprehension. It was so absolutely devastating to think of the minute size of all these mighty vessels anchored there; insignificant and fragile in this awesome, towering mass of storm. At the base of this oncoming monster there was a frenzy of frothing sea, foothills to the mountains. We could see it advancing; it roiled toward us. Higher than we, this boiling spray was a continuous waterspout.

It took no urging to get the clothes off the line, to get all the hatches and ports dogged down, everything double-lashed. We were all inside when the fury rolled over us, first the foaming foothills then the mountains. We wouldn't have dared be out in that. The ship heeled over and snapped on the anchor chain. Then that same horror, the same wind and screaming bedlam. This time we were anchored in the lagoon. I didn't have those awful decisions to make. We knew the drill. We'd made it before and we knew it would end; it was even predictable. That was wonderful.

After a day, the screaming in the rigging dropped in pitch. And we could walk in a stooped position. The barometer started up. The second day we could pick up our duties and become part of the Navy again. We saw the wreckage on the larger ships and the devastation ashore. The small boats all over the beach, the palm trees down, the buildings canted.

It was a severe test of that crew and ship. And we all had this honest pride that once again we'd survived. It sort of bonded us further. And the jokes emerged and the scuttlebutt; all the individual stories and all the unbelievable tales were being stored up. And that's my story of this particular violent interlude: we were "in peril on the sea."

The End, Almost

Kamikaze—Last Act

August of 1945 was typical of those days. This was marking time in the Pacific war. We had patrol duty at Ulithi. We were at sea; we were at anchor. We had a few Red Alerts. No sub contacts. We had spent months the previous year chasing the war. We found it at New Year's and we lost it here. Our concerns were for the next episode of the war. The biggest and the last push, Operation Olympic, the invasion of Japan. So we did not feel left behind. We were just getting the ship repaired and the crew rested. We were ready for Japan. But boring, boring.

The log for 6th of August, 1945:

> Anchored as before in small craft anchorage near Azor Island, Ulithi Islands, with 75 fathoms of chain.

This was our third day at anchor after coming in from routine anti-submarine patrol. The ship was dead in the water. No activity. Sixth of August, 1945. After lunch I went to the chart house. Two guys were sitting on the stools there talking and laughing about something private. I did not intimidate them; I was just the Skipper coming up, getting some stuff. The radio was loud. But I was shocked at what I could hear, even over their laughing conversation.

"This just in. The Army Air Force has announced the bombing of Japan, of the city of Hiro-Hiroshima with an . . . atomic device equivalent to 20,000 tons of TNT."

"Quiet you guys! QUIET!"

They were stunned at the Skipper's unusual tone. The announcement was repeated. I was stupefied! They looked at me strangely.

"Don't you know what he said?"

They could recognize my serious voice.

"The war will END.

You will LIVE.

You will go HOME and the world will NEVER be the same."

I recall those exact words. They gawked open-mouthed. I rushed to the wardroom with the news. The news spread to every corner.

On 9 August Nagasaki disappeared. We were out on patrol again when that came in. We returned on the 14th of August to our anchorage.

The log of the 15th of August, 1945:

> **0000.** Anchored as before near Azor Island in small craft anchorage, Ulithi Island, with 75 fathoms of chain.
> **0902.** Word received to cease hostilities against Japan.

That was at 0902. What about 0903, one minute later? Nothing happened! No screaming, no hollering. A few ships in the anchorage tooted whistles, some shot tracers. I looked around at the other ships. They were all still anchored, but I noticed a lot of men out on deck and we were all out on deck too. Almost nobody left inside; we had to be out on deck. And I talked with so many of the guys. I walked around and there was a little group here, a little group there. And they all grinned and . . . it was just ordinary talk. But it was talk in a new way. The tension was removed from the faces of these guys, the young guys, the older guys. There were more smiles than laughter. And in the wardroom we slowly sipped from the V-J Day Canadian Club bottle that I'd gotten in Panama and hidden in my locker for this day.

My own feelings? I noticed a change that morning. The important change in my attitude came from a relief that would not go away. Operation Olympic would have killed so many of us. And sweeps would have gotten it first, off the shores of Japan. We were glad for the atom bombs. Secondly, my thoughts turned to home and Diana and our future. The pictures on my bunk bulkhead were of Diana. These were more precious now: the woman of honey and silk and spice, and the rod of steel in there somewhere. She had been such a brick all these years. My thoughts moved off the ship. I had been living a life of ship and sea, only. Now I thought of graduate school and a future as a biologist. This ship, this military regimen, was not me really. It was suddenly so very different.

The Navy immediately warned us to stay on full war-time alert, sensibly; the treaty was not signed. The crew understood and they responded with ultimate mistrust of the enemy, so well earned.

The Navy did issue a point system for going home. And it was based on several sensible criteria: length of service, time at sea, combat time and marital status (dependents). I had lots of points and home loomed closer and closer. However, we went back out on patrol after a couple more days. But not until we flooded the Post Office ship with letters home! We could say where we were.

Then on the 24th of August, many days later, this message:

> **1800.** Under way to proceed immediately to Peleliu, Palau Islands on orders CTU 94.6.2 for duty with 94.6.1. Acknowledge.

It seems that Palau had only one YMS and needed another one immediately. Us. We were a minesweeper and the mines up there were uninformed of peace. The war was only sort of over. Despite all my going home points we were heading in harm's way. And to fight such a dumb weapon, the mine. The mine killed friend or foe and saint or sinner; it killed anybody who touched it. And the question was, "Would we be sweeping our own mines, or the enemy's?" We survived the *kamikazes* and we survived the storms but the grim curtain came down on us again. Would we survive the mines of peace? The euphoria vaporized. So the war was not really over for us. Just a different enemy.

Epilogue: Mindless Menace of Mines; Liberation of Palau

PALAU ISLANDS, SOUTHERN HALF

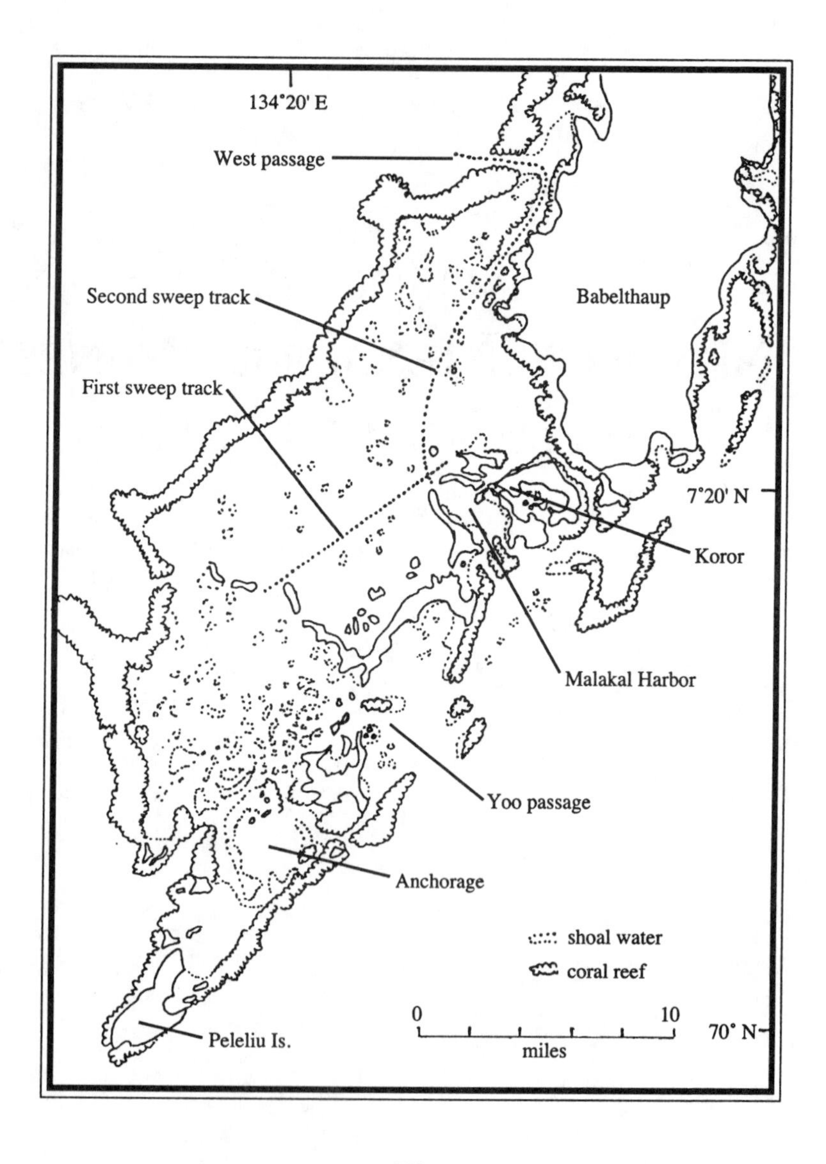

My very first destination when I could get ashore in the Palau Islands was the Port Director on Peleliu Island. The commodore who expected my presence was second. This was the island of such bloodshed a year before and yet the day I was there it was hot and dusty and strangely quiet. I'd hitched a boat ride into Peleliu Island from the anchorage. It was a long narrow channel that could handle small boats only. Ashore there was a jeep going by and I thumbed a ride the rest of the way in. A hellish ride; I was holding on with two white knuckled hands. No seat belts then you know. Quonset huts clustered around the signal tower. We stopped at the Port Director's large Quonset hut; my jeep roared away.

I went in and looked around; a Chief Quartermaster smiled and greeted me from behind a long counter. The bulkhead behind him was a solid rack of chart drawers. I introduced myself and asked, "Chief, do you have a chart of the local anchorage. I came in yesterday and I would not want that to happen to my worst enemy!"

"Are you the YMS came in yesterday?" He wanted to know. "Where are you? We never heard where you got."

I replied in a tired tone, "Why, we're tied up to the YMS out there in the anchorage."

"You're in Shonian anchorage?"

"I guess so, if that's what you call it."

He rushed from behind the counter, which was hinged like a small town bar. Then he turned to me with his hand extended. He called out over his shoulder to the guys in the other rooms nearby, and his voice carried, "I want you guys to meet the Skipper of the YMS that come in yesterday. I want you guys to meet the first real sailor we've seen in this dump." Heads popped out; more grins. I was stupefied. He continued, "Sir, no one has ever come through the reef to the anchorage without a pilot. We thought maybe you dropped the hook inside the passage. Congratulations, Sir, and yes I do have a chart with the best routes marked."

I left in a sense of stupor. I went on to visit the commodore, nearby. And with every step my pride sort of settled in and it was kind of nice plodding along with the rolled up chart under my arm, the red rubber band squeaking. It had indeed seemed a tough passage, but I never dreamed to ask for a pilot. Didn't know such a thing existed in the Navy.

I trudged along the crushed coral road remembering yesterday in detail.

We had arrived at the islands due east of the tower on Peleliu. We gave them the old "dah-dih-dah-dih-dah" on the light and they acknowledged and we were logged in. Then we kind of hung around there, slowly idling. There were no more instructions so we just headed north looking for a pass in the reef. We could see, about ten miles north of us, a cluster of masts. And the radar screen suggested that these were ships but in a lot of clutter on the screen.

So we crept north along the outside of the barrier reef, out on the sea, until we were abeam of these masts and they were indeed a group of relatively small stuff. We could see a YMS, a small floating dry dock, some tugs, patrol boats and LCI's (Landing Craft Infantry). Obviously this was our home. But how the hell do you get there? We had a rather coarse scale chart; it showed a passage through the reef, but my God it was much too small and didn't seem to go through. It was labeled a pass, however probably for native boats. We could see that farther north, about another five miles, our chart showed a Yoo Passage (I don't know how it got named). But it didn't seem to go anywhere except into a coral-studded lagoon. The area around the anchorage was all confused reef according to the chart. There didn't seem to be a clear channel into the anchorage.

Our chart did show that there was an area of about ten by ten miles which was a reef flat. A hundred square miles of the lagoon inside the barrier reef was solidly flecked with reef flats. The chart listed water depth, one-quarter fathom (two feet). And there were little channels here and little channels there of four fathom water. In other words there were a lot of channels in the reef flats but it was basically a solid ten by ten mile piece of reef. A labyrinth.

Well, we knew about coral lagoons but had never seen anything like this! The chart showed a clear area which was the anchorage, twelve to eighteen fathoms deeps. It was about two by two miles; nice big anchorage with a great depth for anchoring and that was where we could see the ships. But how the hell to get through miles of tortuous gaps in devastating reefs.

I mean . . . two feet? Tear us apart! We had to try; hell, those other characters in there had made it.

The first hurdle was a very narrow passage through the outer barrier reef and between some small islands which we had to pass. This was the way into to the lagoon. And the tide was racing out; it was boiling out. But that was good. This was just the right time. It's better to steam against the current because you then have steerageway, even when you're moving slowly over the bottom. The ship would respond well to the rudders with the streaming of that current. To get in through Yoo passage we had to go full ahead just to cut through that boiling mass of water tearing out to sea.

Inside the inner passage we were in a no man's land of reefs. In the distance, both to the north and south we could see large and small islands; we could see all colors of water. The very pale blue was over shallow white sand; deeper channels were deeper blue, and the dark masses were living coral. There were countless coral heads, just sort of lumps sticking up. Some of these shoals were a mile across. Some were ten feet.

We could see those ships anchored about five miles away. There had to be a way! Suddenly we found a buoy, not marked, in one of these channels. I made the quick decision; the only buoy that we'd seen, this had to be somehow a clue to the route south. If I was wrong we were in serious trouble. You can't back out with a tidal current.

Then began the sweat box for which we were now famous in the Port Director's shack. I went up on the flying bridge; you could see better up there. And the Exec went with me and he called down commands to the helm. We had our best helmsman and our best throttle man on the bridge. Some places we could go five knots, which is about a fast walk. Other places we were just creeping. At such times it meant steering with throttles, and the two screws, port and starboard screws. And we'd go "Half ahead port, stop starboard. Rudder 'midships. Hard left, stop port. All stop." Then we'd coast. On we went this way. There were helm commands every minute. We just missed some of those reefs by ten feet. I could feel the stern slipping and sense that we were going to whack the reef. I gave the throttle order to increase one side or the other, to swing the stern away from those things. Had to do it with the screws, quickly.

We came on another buoy about halfway. "Hey, we're on the right track." The buoy didn't say anything; it wasn't red, it wasn't green, didn't have stripes. We sure hoped it wasn't a blind end, we could not possibly have turned around.

The narrow channels that we were going through were actually formed by tidal waters running in and out. And at flood tide the waters flowed into places like the anchorage. When the tide went out all that water had to rush out and that's what was happening now, a strong current out. But the centers of the channels were safe. In tidal channels, live coral heads can't grow, because the larvae of corals can't settle in the strong current. So we didn't expect to find live coral in the middle of these tidal channels.

We made some right angle turns. We made even some double-back turns. We had to have that current against us or we would never have done it. If we had water flowing from behind us we would have been swung right onto these coral heads.

Another buoy and we knew then that this had to be a marked channel. Little buoys, miles apart, not on the chart. It looked like the widest pass was opening up, seemed to be heading toward the anchorage, and it was. Finally, a couple of hours later, we had eased out into the open anchorage where we lost our tidal current. There was a gentle wind, calm water. Everyone slumped. There weren't many people who were unaware of what was going on all the way in. We blinked over to the YMS, "dah-dih-dah-dih-dah;" we eased alongside them and tied up.

The Captain on the other YMS was a J.G., junior grade lieutenant. His old skipper had just left. And he was so glad to see a full Lieutenant. Suddenly I was senior minesweeping officer of a fleet of two YMS's. Here was a switch. All these years, all these months I had been in a column or an escort of some convoy; Number 3 position. Follow orders. One tooth on one cog of the big machine. Now, I was in charge of the sweep program. I developed the notion that has stood me in good stead since. If I didn't do it some ass would.

We shot the breeze with the officers of the other ship. They had swept around the anchorage but had not gone farther north. The lower quarter of this archipelago was all that had been used. The rest was in enemy hands. No one had gone up there. So they hadn't done any sweeping beyond the southern region. My head was swimming with plans. I studied the charts.

And the next morning I went in, and I've told you about that adventure in the Port Director's shack. And I left you with my walking towards the Commodore's office. This dusty way took me to headquarters, Island Command, one star blue flag in front of it. A truck passed, dust; a plane landed nearby, roar; this was near the airfield. Heat shimmered.

I was greeted cordially by the Commodore, but he got to the point quickly, no small talk. The route to Koror and Malakal Islands in the middle of the archipelago, where all the Japanese were "withering on the vine" (the MacArthur technique) had to be cleared of mines immediately. The pass through the reef also had to be cleared immediately. Word had come in that a Japanese ship was starting on its way from Japan to pick up the troops at Koror. We looked at some charts. Tough prospect.

Then an amazing, preposterous proposition came full-blown from my lips. I didn't believe it when I thought about it later. I said, "Sir, I have swept for invasions, and swept in the Atlantic, and we had to plow into unswept waters, but I don't think we should do that in peace time. There are three steps to take before we take the ship up there to sweep." I listed them. First a day to reconnoiter. I wanted to go up there in a small wooden landing craft that would not set off a magnetic mine. I had in mind an LCVP (Landing Craft—Vehicle or Personnel). It drew only about a foot. Second I needed to know anything that could be found out about our own mine fields. Did we mine here? I thought we had. I thought I'd remembered that in the news, or in the Office of Naval Intelligence reports. And most of all I wanted to know what the self-destruct rate was on those mines. When would they be disarmed? And third I wanted to interrogate some Japanese naval officer in the area, someone who knew about their mine fields.

I think he was somewhat taken aback by this tough talk from a kid Reserve officer. "How much time?"

I said, "three days."

He didn't hesitate, "I'll assign a LCVP and a coxswain. I'll try to get a Japanese naval officer. And I'll wire Pearl about our mines. Get going. Report here in three days."

"Aye, aye, sir?"

I did an about face, chart in one hand, my cap smartly tucked under the other arm.

That was the start of two rather bad months of sweeping. But in a fascinating group of islands. The most beautiful islands in the world. I swear. And I was going to be part of liberation and restoration after a war.

A war that had savaged this place and its people. We had invaded Peleliu and really only obtained an airfield. There was no anchorage here for even a destroyer. With only the airfield, we neutralized the rest of the archipelago. There were spectacular harbors up in the middle of the islands, and deep water passes. The enemy had large parts of their fleet here at one time, but our air presence denied that. The enemy was regularly bombed and they had indeed "withered on the vine." And now they wanted me to open up the middle islands and that long deep passage out to sea and bring the end of war to this beautiful place.

This archipelago sits on a platform like most oceanic island groups. This platform is about fifty by twenty miles, roughly north to south, at an angle. Most of the platform is submerged and ringed by an active coral reef on all sides. For the most part these are barrier reefs. The barrier reefs are ones that are separated from the islands and enclose a protected lagoon.

But these were high islands, not a low atoll like Ulithi. And the longitudinal spine of this platform is a row of mountains forming a few large islands (one over ten miles long) and scattered small islands; very numerous. Along the island shores there's usually a fringing coral reef growing onto the hard rock. All over there were extensive coral reef flats, very shallow, with a maze of twisting tidal channels, like those that gave us the trouble on the first day coming in. On the west side of the central islands there was a huge lagoon, as wide as ten miles but with only one really good pass through the reef for a large ship. That made a fantastic anchorage, so protected. The central area was densest in population, and the islands of Koror and Malakal were the commercial center. There had been a copra trade and coal depot. And it was the Japanese military center, so the central islands were important, and this was what I'd been assigned to get through to. To make available. The northern part of the archipelago was the high mountainous island of Babelthuap, covered by rain forest; here were most of the native people of Palau, cousins of my Ulithi friends.

The islands are very beautiful, bright tropical green. And so many of the smaller ones seem to be eroded remnants. And these have the shape that is special to these islands, as far as I know. They are hemispheres on an undercut base. Very striking; mushroom-shaped. The waters are coral dominated everywhere, every color of blue, from a rich clear blue to very pale and pale blue-green. Nowhere more beautiful.

So, as I promised the Commodore, my first job was to make the trip none of our people had ever made. Destination: Malakal Harbor and Koror, the heart of the enemy fortifications. This started with a retrace of the tortuous trail through reef. The LCVP barge showed up, and I stood at the ramp forward (the ramp which could be lowered for the troops running ashore); I looked over the ramp down into the water, and we moved slowly. Very clear water. We finally came to the large lagoon anchorage and headed north.

Every mile we dropped a Dan buoy; this is a minesweeper's buoy made of yellow balsa wood with a pole and flag. We made a straight line to the opening into Malakal Harbor from the pass below. This was a long trip, ten miles or so. I searched for mines. The buoyed lane was for other ships to follow; they would be able to guide on those Dan buoys, in waters where I had not seen a mine.

I was also terribly aware that we had no word on whether the cessation of hostilities had been heard up here. We had broadcast the Emperor's surrender message, in Japanese. But did they get the word, and would they respond to it? Or would they shoot? Down in the bottom of the boat was my gunner's mate. He was a tough guy; I liked him. And he had a tommy gun, out of sight. I was in full tropical uniform without side arms. I would not take a gun. I had figured that this called for someone who looked so secure that he wouldn't need to wear a gun.

We saw a big cave and bigger gun facing us from the Malakal Harbor entrance. It covered the harbor opening—spectacular! Were they in there? I saw no mines in that ten mile line. Seemed odd except that this was a straight shot from the Peleliu area up to this entrance of Malakal Harbor. Maybe they didn't mine there because, after all, before we came they must have made that trip down to Peleliu. They had built the airfield down there.

Slowly, we entered the harbor and the land all around swallowed us. The gunner was pouring sweat; he couldn't see where we were going. Malakal Island had a spectacular quay with a rock face and deep water alongside. It was huge. It extended out from the island, I would say a couple of hundred yards, and probably more than that wide. Several ships our size could tie up. And I could see Koror, the city and island adjacent to the north. A causeway connected Malakal and Koror.

No one shot. I saw no one. I'm sure they were peering at me as we tied up to the quay. I got out and walked along the quay. I didn't expect land mines. It was a mess, a shambles, all kinds of bomb craters, destroyed buildings and bombed out oil tanks. But the tropical vegetation had taken over. The docking area obviously hadn't been used for a long time. I saw no mines in the water alongside the quay; I walked up and down. But then that makes sense. Why would they mine their own quay? I had a lead line for taking soundings and we had six fathoms alongside the quay—great. I wrote down some notes; reboarded and we retraced our path. I did not look back—I was consciously on a stage and wanted to look absolutely sure of myself. It was the most . . . uneasy exit! No mines in sight on the way back. We followed the line of buoys that we'd planted on the way in, this time on the other side.

Our foray took many hours. It was sure a lot easier making that run up those twisting reef channels in an LCVP than it was in the YMS. This guy knew the way. He just rolled it this way and that way and we swung in and out, leaning over and shooting ahead. We plowed up to the YMS nest. Everything seemed so ordinary. The day had not been ordinary!

We made plans; studied the charts and I just got consumed; lived up in the chart house. I conferred with the officers. The other YMS skipper agreed on the plans and I radioed Peleliu of my finding no resistance and that I had officially "taken occupation." It kept me awake a bit that night. After all, those islands which I had visited had been closed to outsiders for a generation. Forbidden islands. And though I had not seen the enemy, they were there.

No outsider had been allowed in here since Japan got these islands after World War I as mandated islands. They illegally fortified them and I recalled a book by Willard Price called, *Japan's Islands of Mystery*. I think in the '30s before the war, he managed to get in there. I can't find the book; I'd like to have that. These surely were one of Japan's islands of mystery. They had fortified many islands all over the western Pacific.

The commodore called me to a conference the next day and he had the Navy data on U. S. magnetic mines. Most of our mines had saturated Malakal Harbor where we'd just sailed over in the LCVP. A wooden boat thank God. The self-destruct time was this month. Ouch! How does a mine self-destruct? A thin piece of metal encloses the mechanism. This metal, in this salinity, in this temperature, corrodes at a known rate. Then it perforates and floods the works; the mine is dead. The explosive isn't damaged but the arming mechanism is ruined. Now, was I going to assume that they had all drowned the month they were supposed to—by the thickness of a piece of metal? No way.

I told the Island Commander I would have to do a magnetic sweep in Malakal Harbor and I did not want to take the YMS in there. Though we were a wooden ship we had lots of metal and we had not been "degaussed" that I could remember, ever. In other words, neutralized magnetically. He also told me that intelligence was dickering with the enemy for a mine officer. Great.

I asked if we could sweep tomorrow. First I wanted to go for moored mines along the buoyed lane. Wanted to go along each side to make a good mine free corridor up to Malakal Harbor. Then later we'd do magnetic and moored mine sweeps into Malakal Harbor. I had seen several wrecks in shallow water; it had been mined all right, and effectively. I had an idea and asked, "Could we have the LCVP again tomorrow?" By the third day I'd promised him I would establish our presence by tying up on Malakal Quay. Could I have an interpreter? The Commodore nodded yes for everything. He seemed pleased and I took off. "Thank you sir."

We were back in combat, the daily terror of the unseen mine. Our sweeping started on that track towards Malakal Island in the western lagoon. We swept one side, the starboard side, of that line of buoys, just grazing them. The ship was actually in the track that I took earlier in the LCVP. I felt very certain that we were not going to get a mine, because I'd searched that water. We swept, but we got no mines. We swept back, and had a two hundred yard wide channel now, with a row of buoys right down the middle. We did this twice. The main channel was clear.

But what about the U. S. magnetic mines in Malakal Harbor? Since I didn't want to take the ship in, we used the LCVP. The ship sat in the harbor entrance and the LCVP hauled our cable in, our magnetic cable. It was more than a quarter of a mile long, hard for the landing barge to do this. When the LCVP got back alongside of us, we just sat there and pulsed the juice out on those cables, setting up a giant underwater water battery. No mines. Over and over we pulsed, many times. These magnetic mines could be on a ratchet set to blow on the twentieth stimulus. We probed the cable in all directions all over the bay, always keeping the YMS in swept waters. We worked the next day. No magnetic mines. They had indeed self-destroyed and we could only thank God for the good craftsmanship in the U. S. of A. ammunition factories.

We then swept for moored mines all over the bay and found none. We examined close to shore from the LCVP. Nothing. We were in a harbor that had no mines. I was convinced of that. Or if . . . a mine were there it was in some corner and we hadn't found it.

We pulled up to the big quay and radioed Peleliu, "Mine-free lane to Malakal and a mine-free harbor." The next day another ship was sent up, an LCI; now she could come in swept water all the way. And they could be a mine destruction vessel for us; they would come along and shoot them up as we swept them.

Next was the passage through the reef out to sea. A U.S. Army interpreter came out and a Japanese Navy warrant officer hiked out from Koror town. Don't know how he got over the broken causeway bridge. Our interpreter was very good, young guy, Japanese speaking Californian. The enemy warrant officer told us there were a thousand moored, contact mines around the islands. Mostly around the approaches to the central area. Earlier, the U. S. minesweeps had swept a lot of mines off the coast of Peleliu during the invasion. Some sweeps were sunk. These central islands had been mined against our invasion forces, and possibly even quite shallow mines to get smaller ships. In the West Passage, going through the outer reef, he said there were twenty mines. Somehow I believed him. He was actually a very surly guy; the interpreter said so too. Why not? This was no fun for him. On the other hand he was not going to be sweeping the mines so maybe it was in their best interest for us to know exactly where the mines were.

But this was a new ball game now. I would not take the ship through a narrow passage with twenty mines in it. I talked things over with the other officers. We had a plan, largely through the work of our wonderful engineering officer, the scrounger. I radioed that we wanted an LCM; this was a landing craft which could hold a big tank. It was fifty feet long, a large barge with a powerful engine. It only drew two feet; a very shallow draft allowing it to run up on a beach. We made this LCM into a minesweeper! The engineering officer was a dynamo. This was his baby.

It took a day, after our crew set to work. You can't believe these ingenious guys we had down in the engine room. They welded and bolted our drums of wire on the little landing barge. They put on davits, these little cranes for swinging out gear. We put on an electrical winch run off the engine. She could barely pull the gear when they streamed it out. They tried it out in Malakal Harbor; it could be done, slowly.

We got an all volunteer crew and our engineering officer and a local mongrel dog. One of my officers went up in a little Piper Cub; we wired down to Peleliu for a Piper Cub! It could fly at less than a hundred miles an hour and they scrutinized the reef passage. They counted nineteen mines in the water, viewed from directly overhead. I was so glad we had not taken the 353 in there. We probably would have hit one of those nineteen.

Our LCM went out with her gear streaming, and a little ensign snapping. Extraordinary job. They got all nineteen! And the Piper Cub was flying over and there was chatter: "Baby Girl, Baby Girl, this is Baby Boy. Do you see that mine dead ahead of you?" The barge went back and forth several times after they had gotten the nineteen. No more mines.

Baby Girl returned triumphantly and we set out in the YMS for the final check sweep. The 353 headed out, gear streaming smartly. My God, we got the twentieth mine! Nerve racking! Imagine, you think you have all the mines and you get one more. The ship missed it, the sweep gear got it. Good decision, not taking the 353 first.

Our LCM minesweeper made Naval history. This was our baby and she worked where we could not. The volunteer crew wore the elite mantle. I have read since that the last mines in the waters of Palau were scoured out by "a landing barge."

I radioed Peleliu, "Koror and Malakal Island have an open lane to the sea. Traffic anytime. Have pilot ready to follow the swept lanes." We tied up at the quay. Guards were posted. The LCM tied up next to us. We were all proud. We truly were proud! The sensation was so special. For years we fought but now we had liberated, opened the entrance to the world for Palau.

We did see some Japanese. They were too curious not to peek. And we saw some native Palauans; some men made the trip in an old power boat. Never knew how they felt about all this. Somehow we felt at home by now on Malakal. We had explored and cleared the water. By God, this was our own bay! We were smug. But 980 mines to go! The worst was not over for us. There was no enemy, but we still had a left over war.

A remarkable event happened the first morning we were moored alongside the quay at Malakal Island.

The gangway watch came to get me, and I went out on deck; he pointed. He pointed towards Koror town. And there, slowly, winding down from the causeway to Koror was a formation of men. Looked like maybe a hundred. They were in fact marching in columns and they had a British-style arm swing, that exaggerated parade ground march. They had a military bearing, eyes forward, no smiles. They were led by about four officers. This was a small army. They were in ranks.

Then I realized that these were very tall soldiers, from India. They had turbans, or attempts at turbans. All over six feet tall. They were in fact the Imperial Singapore Guards, an elite unit captured four years ago and held here as POWs. They were completely emaciated. I don't think most of those guys weighed a hundred pounds. They came on, marching to a silent metronome and an occasional bark of a non-com. The precision of the march eroded at the tail of the column: here men were staggering, crutches, canes. They were dragged; they were carried; they were clinging to life, and to honor.

No one had any idea these POW's were here at Palau. I was reminded, of so many years later seeing the movie, "Bridge Over the River Kwai," and the rag-tag British Army marching to the bridge site. Same cadence, rags, and bearing.

This unit marched up to our ship in brave malaise; they halted on the quay, and they formed up. Pitiful to see the stragglers pull up. The commands were barked. They came to ramrod attention and saluted the colors of the YMS 353. I had had the intuition to put on a shirt and cap, thank God. I returned the salute, and asked the major (he was in a remarkably preserved uniform) to stand the men at ease and fall out. There was no shade here at exactly seven degrees north of the equator.

When I first saw these ranks approaching and recognized them for military POW's, I ran up to the radio shack and wired Peleliu. Then ran into the galley and told the cook to make soup. "Make soup for a hundred, no pork. Quickly." And all the tea that he could make. And the Exec organized the crew to gather up all the edibles from fruit to bread, boxes of cereal, candy, everything. We looted our lockers.

I invited the major into the wardroom. And we heard the gory and malevolent details, spoken with eyes filmed with tears. Again, the sick Jap view of prisoners. These were not soldiers; these were not men. So they were not fed; they were not clothed; they were not sheltered. They lived in a compound and they ate mollusks from the sea, anything they could catch. They boiled grass leaves and algae for soup. Almost all were sick. So many died. No medical care; this was very apparent to us just looking at them.

I went back up to the radio called Peleliu again, suggested that an LCI be sent up to transport these people to the hospital. I said "They're in terrible shape," and suggested sending a doctor and a crew from the hospital. Our pharmacist's mate was giving first aid. Pitiful. Morphine for the dying. These silent bundles of men, just sat there: slowly chewed and sipped. We had soup in cups and bowls, we had soup in glasses. And bread. And tea. And sugar.

The major came in to join us at lunch and he savored a slice of bread with butter and a real cup of tea, which he caressed in two hands. And there were potatoes and roast beef. More tea, ice cream. He was obviously overcome. Can't you imagine; why not, after four years of living like a dog? What an event. I can't forget one single item.

An interpreter, a doctor and an Army Captain of Ordnance joined me and my officers on an inspection tour of Malakal and over to Koror. They were going to report back to Peleliu on the health and military problems.

By now it was true peace. On the 2nd of September, the signing on the deck of the *Missouri* in Tokyo Bay took place; there was no longer the possibility that we were in enemy territory. The Japanese we saw were pitiable. They were cowed; they were exhausted; they were unbelieving; actually they were empty. For a year we had closed their port, took their airport, bombed them regularly. There was no city, no defense facilities. The developed area was burned wreckage, the shells of boat yards and shops, the shells of homes; no usable streets or roads, no electric wires. Several ships in the harbor were protruding from shallow water; they had been beached there. Others were below water. Dozens? War for the last year was devastation. Complete attrition, withering on the vine.

All they could do was hide in caves, which riddled the hills. I recall mentioning that we saw a cave looking out over the entrance to Malakal Harbor when we went in on that LCVP. We went up there, the Army captain of Ordnance and I. There was indeed a huge naval cannon there, the one I'd seen from the bay. But it was awry; it had apparently been exploded. That was the sign of a beaten army of course. But we waded through a knee-deep mess of what looked like dry spaghetti. The captain said, "Smokeless powder. See." He lit one strand with a Zippo lighter and fizz, My God! I was horrified. I think he was looking for my reaction.

We saw all the underground fuel tanks. Demolished; they'd been direct hits, gaping holes on Malakal Mountain. A sea plane ramp at the base of the Koror Causeway sticks in my mind. Two reasons. First, the Captain was along and there were some land mines right at the water's edge. He kicked them! Then he assured me they were too corroded to work. I saw the dredgings for the sea plane ramp. The dredgings to one side had two huge shells, out in the open, up on dry land; these were the giant clam, Genus *Tridactna*. Each of the shells was the size of a bathtub.

I rowed out to some of the wrecks. Most had only the bridge and the mast out of water. The rest, probably with the bones of the Japanese crew, were underwater. Above water everything was ghostly. No people. Reflection of sunlight on water played on the bulkhead. The gentle lap-lap of water. And the bridge had . . . papers, books. There was the wheel and the engine room telegraphs and the captain's chair. Ghostly. Nothing was booby trapped, just deserted, stilled, dead.

One cave was an Officers' Club. A shambles, it was abandoned. We could smell Oriental food and stale beer. And all the little things: bowls and little glasses, and rumpled rugs around. Low tables. There were shelves with some books falling over. Drawers with paper and pencils and cards. And condoms! One drawer, it was filled with condoms neatly wrapped.

The officers were not without company. We found three girls, teenagers. According to our interpreter, they were brought down here to be companions and servants to the officers. They smiled. Their little shack of tin sheets to keep the water off had a huge metal pot in front. It was full of cigarettes. They said the guys from the SeaBee group that arrived the day before had spent the night here. Gave them five cigarettes for a place in the line. Hundreds of cigarettes! They said they were busy all night, grinning. The doctor was astonished. "All night?"

"Oh, yes." And they had graphic means of describing the SeaBees.

The doctor was concerned. He said, "Come over with me." He motioned them over behind the bushes to examine them. They returned and he grinned. "Not a scratch. Impossible, not a scratch!" Then, "They thought the exam with instruments was a new kind of come on." The girls, I gathered from the interpreter, were probably Korean. They were young and attractive. They were not emaciated.

The most satisfying sight was the clean up and restoration. The SeaBees were outstanding! The Marines landed and the quay became immaculate; a parade ground around a big flag pole was built. It was lined by white painted rocks. Some of the native people drifted in and began working. We brought shiploads of food and medicine, clothes. The Navy worked at peace and did darn well.

Finally the Jap destroyer came in, a troop carrier. She came down our channel, uneventfully. And there was a ceremony of departing troops. They marched onto the quay and actually laid down their arms in huge piles. They formed up, saluted and boarded the ship. The U. S. Navy was not involved. This was strictly a Japanese show. We watched.

Something changed as they disappeared. As they steamed slowly out of the harbor, the work of reconstruction seemed to accelerate. A weight seemed lifted; and it was. Even though the Japanese themselves had been defeated they were the hated, occupying enemy. For Palau, a generation of occupation.

We had so many hundreds of mines yet to clear. It seemed strange that a recent college kid like myself was the senior sweep officer, now forced to take the lead. I had never had any formal sweep training, not even the Yorktown Mine Warfare School. I just went aboard and worked on that YMS out of Norfolk. So now I had to study and plan on my own, but of course, with the constant help of the other officers.

The obvious first step, after that first sweep of the track to Malakal Harbor, was widening it. We did that. We never stopped sweeping, both magnetic and moored sweeps, mostly moored sweeps. And we began to get more mines. Nothing as dense as the twenty that were packed in the west passage through the reef.

The huge western lagoon was the first objective; I wanted to clear that whole lagoon. Our Japanese informant said they had mines there; they had laid mine fields to protect this deep water anchorage for their fleet. This western lagoon I think they called "Komebail" Lagoon. That's the lagoon that had the coral reef pass. A huge anchorage, ten, thirty fathoms of water; it led right into the commercial center, Malakal Harbor. I studied the chart and picked sub-areas to sweep, marked them out on the chart. We located them by lining up islands; we had these marked on the map, and we could find them out there. There were not enough buoys to mark them all. So we had the sweep plan on the chart. Weeks of work ahead.

These areas were tackled one at a time. We wanted to sweep water of a given depth and we needed straight lines for our lanes. Hopefully we could sweep it all at a depth that would make the water safe for deep draft vessels, for a carrier, or big troop transports. In tough little corners and shallows, and little pockets here and there, we could send in our little LCM sweep.

I personally checked out the tough little places by going over them in the little LCVP, which did a wonderful job. And I went swimming with a mask where a boat could not venture. I walked around on the reef flats at low tide and I checked for mines. However, there was a little time to study the reef, as a biologist of course. Came back to the ship stimulated, tired, and ready for a shower.

Strange, at a time when I had this heavy responsibility and very real danger, I felt more and more like a graduate student in biology. A guy with all those stateside points was still out here. One by one the officers and the crew left. Not the skipper.

September turned to October. Out sweeping by day; Malakal Harbor at night. Sometimes we anchored in the lagoon at night to get started earlier. We made lots of runs south to the anchorage at Peleliu. That was still our home for mail and supplies. The Peleliu headquarters were there. That was really the home port. We spent some days in floating dry dock. Old stuff; boring, boring. And then out sweeping; the same old stuff. Week after week of minesweeping was routine. Dangerous but dull.

The only excitement came when we shot a horn off the mine and the damn thing would bust into bits. Blam! We also had excitement when we lost our gear. We'd hit coral heads and the cable would SNAP, and ZING would go the sweep gear. Even the crew was intrigued by this stuff that came up on the cutters, like clams, an octopus, and a bunch of black coral one time.

Having black coral hung up on the cable was exciting to me. It was also exciting when a mine got hung up on the cable. We were used to these "hang ups;" but one time the cable acted very differently. It slanted down like we could have a mine on the cable, snarled in the cutter, but under the stern of the ship. Not coral or clams. A mine! The bosun called me; I went down and we reeled in the cable very slowly. What would make all my "go home points" kind of pointless would be pulling a mine right up to the stern of the ship. We couldn't see anything, but the wire angle was down too steep. This pointed to a mine under the ship. We didn't have much cable left out. I had the winch stopped right away, then stopped the ship. Now we were sitting there drifting with the mine maybe very close, maybe under our stern. We could not see one.

On went my mask and off went my shoes, and I slipped over the side. I had no idea how huge a mine would look underwater! I said to that mine, "Please don't cut free and pop up under the stern." If it did I'd never know what happened. I was just a few feet from that damn thing. So I popped up again after I had seen enough of that thing to know where it was. The danger was very clear. I hollered up to the bosun, "Take the cable drum brake off." And added, "Get all hands forward up on the focsle including the engine room gang. Everybody! Nobody under an overhead. Out in the open on the focsle." Nobody needed to be told twice. They just disappeared like magic. I could see this mob of guys up on the focsle staring down at me in the water. Sober.

Then I hollered to the bridge from the water, "Take her ahead slow. Very slow!" And with the brake off the cable drum, this should pay out the cable. I dove again and I saw the screws turning slowly; we were pulling away and the mine was pulling the sweep cable out astern. Finally it cleared the stern and I popped up again. The ship was almost drifting in this slow pace, so I swam alongside. I called for a rope to climb back. There were more hands than could reach me, trying to bring me aboard.

We pushed the throttles ahead and ran out a couple hundred yards of cable; then we threw the brake on; zing, the cable went rigid and up popped the mine. She just came up big and black and ugly, that terrible menace. But no longer; it was on the surface. And our scoreboard of mines painted on the outside of the pilot house grew by one more hashmark. We couldn't wait to shoot the hell out of that thing. Finally sank it.

What was needed on this job was not dash and daring. What was needed was calculation and care and patience. Attention to detail. That was my natural approach, always had been. Looking back I could see that I was more and more becoming a civilian doing hazardous work. But I could not have done it without the years of ship handling and seamanship. I knew the drill. Knew the sea; knew the coral, knew the ship and knew the crew, now such pros. I had by now acquired a seaman's intuition.

We swept, as before. Every day. The other YMS did too. We were seldom in echelon; that was for wide open water. And the LCI came along as mine destruction vessel, and as rescue for survivors, in case. They were ready. And the master chart got more and more cross-hatching of the areas of swept water at a given depth. We were very careful to start our runs with the ship inside well-swept waters. Like mowing a lawn with one wheel of the lawn mower on the cut grass. We were careful. No swashbuckling; not now.

The uneasiness was unending; sudden death by mine hung over the wardroom. Such feelings are undebatable; words do not function. We planned of course, went over charts, but this pervasive menace was so anachronistic. Everyone else had gone home! The war was over, two months ago.

During the Philippine invasions we could feel the utter madness in the *kamikaze*. But all wars are a form of madness. And now we had the mine as the ultimate madness playing in the theater of the mad. Ashore, we were rebuilding Palau and out here on the lagoon sudden death hovered every day but no finger was pulling the trigger! One officer got his orders for home—pending arrival of a replacement. Each day after that was a gray-faced horror for him. Routine sweep? No. Not for him. The ultimate madness.

One time only were we in what was supposed to be swept water and it was not. An error.

I was on the signal flag deck and the Officer of the Deck was on the flying bridge. He had the best view up there. We were relaxed and were approaching our first run. Our gear was out. The men hadn't yet put their helmets on but they had their jackets on. Some of the engineers were sitting on the starboard rail smoking and sunning and chattering.

Suddenly I heard, "Skipper, hurry!" I knew that tone; knew the urgency. I cleared the three steps up to the flying bridge and saw the lookout. The lookout was already on station up on the bow, and he tried to shout but nothing would come out. He pointed. Ahead, just to starboard: a shallow mine. Too close! We'd take it on the starboard bow. It was just a bit to starboard of dead ahead. Instantly I hollered down the tubes, "Hard right! Stop starboard engine! Port ahead full!"

"Aye, aye, Sir." There was a brittle tone in the helmsman's voice. One of the officers looked at me in disbelief but could not challenge my order. Turning away certainly must have seemed more sensible than turning toward that mine.

Looking back, this was a twisting maneuver in my mind, a maneuver that would swing us around the mine, crabbing sidewise by turning toward it. Skidding again. My intuitive response was the right one; no time for a second choice. But then there was that awful momentum until the screws and the rudder took hold. The demands on the engines shook the ship, we could feel the vibration, but still we plowed ahead towards that mine. Right towards it on the starboard side. The lookout was crouching behind the anchor as though that would help. The bridge was silent, frozen in time. Back on deck the engineers were still laughing. This was all happening too fast. I couldn't holler to warn them. I had entered my emergency trance.

Then we slowly plowed to one side, and as we approached the mine, our stern was swinging away from it. As we turned toward it, we passed it, at the same time. It was only a few feet away all along the starboard side! We stared, horrified. There were the horns and barnacles, the algae. The engineers saw it go by a few feet under their butts on the rail. They flopped down on the deck! And then it was astern. It was wipeout on the bridge.

We had to react again, quickly, to get back on course. She straightened out; that took the slack up out of the cable, so the mine was now in the path of the gear. There was a tug and the mine popped clear of the water and splashed down about a hundred yards behind us. We were our own mine destruction vessel on that one, too! Everyone wanted to shoot; it sank. We finished the area, went home.

"Moored as before. Malakal Quay." We got mine after mine after mine. Time drifted until it was well along in October; on 23 October mail came up on a ship from Peleliu.

A letter for me; my orders! My orders from Administrative Command, Mine Craft, US Pacific Fleet: "Subject: Orders, Release from active duty." Etc., etc, etc.

Impossible to believe! We were tied up, Malakal Harbor. Next day, 24 October 1945, I have this set of orders in my file:

> From Commanding Officer YMS 353
> To: Richard V. Bovbjerg, Lieutenant, D., USNR, 237072.
> Subject: Orders—Release from active duty."
> [The letter was two lines:] "You are properly relieved and detached this date. Carry on your basic orders."

It was signed by my most recent Executive Officer, now Captain of the YMS 353.

We had a brief ceremony with the crew lined up on the focsle. I slowly walked down the line; there were strong ties here. Grown stronger. The handshakes were very genuine; I loved them so. And I worried about them now. How silly! They were good at their job. But I worried about them. I was leaving. How can they get along without me? (The log shows they did just fine!) I went down and packed and the ship's work continued.

In September an "All Nav" had come to all ships directing the commanding officer to designate a ship's historian to write a history of the ship. I had elected to do the job for the YMS 353, and had not finished it when my orders came; the new skipper begged me to do it because he claimed that he simply could not. That absorbed the last night I was aboard. The yeoman typed it after I left. Years later I visited the Naval Operational

Archives in the Washington D.C. Navy Yard. A helpful archivist found our copy and made me one. I think each of you guys has a copy, just a few pages.

The labor of doing that history, short as it was, involved lots of log checking. An outstanding fact emerged; we entered the Pacific on 1 September 1944 and final Peace was 2 September 1945: one year and one day. The number of real combat days was one month and one day. What a ratio! By far the most time was en route or anchored somewhere. That is what we did. Our time spent "smelling burning powder" was so little. Hurry up, wait. shoot.

Next morning we headed down toward the Peleliu anchorage, through that huge lagoon. And there on the edge of that huge reef flat was a rusty, raw steel barge; it was anchored in good water. But it was just to the north of that tortuous trip down to Peleliu. It was used as a pickup station; landing craft could come and pick up somebody or drop off somebody, or gear. It would save a trip through that awful reef. We'd radioed for an eight o'clock pick up but we arrived quite early in the morning. I would go to Peleliu; would log in there and stay at the Bachelor's Officers' Quarters till I could get a flight out.

We pulled alongside the barge; didn't need to tie up. They tossed down my canvas sea bag and I jumped down the few feet. The rusty thing was already warming in the morning sun. And my ship slipped away and headed toward that pass back into that lagoon, "steaming as before." The embarrassed crew sort of smiled and waved. I tried to smile. I did wave. I stood there. The wake of the ship set the barge to rocking a little bit. The officers on the bridge waved and the new Skipper, my old buddy, hollered something that scattered in the wind. Having a buddy is a love like no other; we had shared it all, all the way, he and I, as one. The ship went through the passage; she turned to starboard; she was gone. She went behind Gambadoko Island, and disappeared. She was steaming off as usual. But not I.

I was alone on that lagoon. I was alone in the world I had come to know so well. It was my world; the huge blue sky—very blue, and the sea breeze had a salt tang to it. The water was so clear, and I could see the little fish and the coral heads. They were marked by the turbulence and froth of the tide running in. I couldn't see the barrier reef; that was out of sight. Barely visible were the masts of the ships to the south in the Shonian anchorage. This was my world, so ordinary. And yet here I was, waiting to leave it.

As my old ship disappeared so also did that unending terror of mines. The unending demands of command were gone. And what was my duty now? I couldn't understand a thing. What would happen tomorrow? I drifted into a void.

And suddenly I cried. God, how I cried! I sat on my seabag and sobbed incoherently. Then I realized that in my arms I was clutching a flag, the one the crew had hauled down from the yardarm. I whispered to Diana, "I'm coming home. I'm coming back to school. And we'll have a family." The agony in my throat intensified and choking pain seeped through my whole body. I looked for the ship but she was gone. The rusty red barge rocked. And that hunk of steel and I shared this loneliness. I cursed at the pickup boat for being late even though it wasn't eight in the morning yet.

I studied the flag in my arms. It was sooty from the exhaust fumes and frayed at the edge by the storms. It was riddled with shrapnel tears and it was faded by the tropical sun and the salt spray. But it was my life. I clutched it desperately.

I add two footnotes. Two happy months later, before Christmas, to preserve that flag I had it dry-cleaned. In a couple of days I picked it up. They handed me the box, and it had a note on it. "Thanks. No Charge." No charge. Two years later I came on that box and opened it up, for a quick remembrance. It was a pile of dust and the dried skins of moth maggots.

Geographic Index